W9-CMD-327

THE JAZZ SCENE

THE
JAZZ
SCENE

An Informal History from
New Orleans to 1990

W. Royal Stokes

New York   Oxford
OXFORD UNIVERSITY PRESS
1991

For Erika

Oxford University Press

Oxford New York Toronto
Delhi Bombay Calcutta Madras Karachi
Petaling Jaya Singapore Hong Kong Tokyo
Nairobi Dar es Salaam Cape Town
Melbourne Auckland

and associated companies in
Berlin Ibadan

Copyright © 1991 by W. Royal Stokes

Published by Oxford University Press, Inc.,
200 Madison Avenue, New York, New York 10016

Oxford is a registered trademark of Oxford University Press

All rights reserved. No part of this publication may be reproduced,
stored in a retrieval system, or transmitted, in any form or by any means,
electronic, mechanical, photocopying, recording, or otherwise,
without the prior permission of Oxford University Press.

Library of Congress Cataloging-in-Publication Data
Stokes, W. Royal.
The jazz scene / W. Royal Stokes.
p. cm. Includes index.
ISBN 0-19-505409-1
1. Jazz—History and criticism.
2. Jazz musicians—Interviews.
I. Title.
ML3506.S87 1991 781.65′09—dc20
90-14208

1 2 3 4 5 6 7 8 9

Printed in the United States of America
on acid-free paper

Preface

This book is not a formal history of jazz, although it contains much history. Nor is it an analytical or critical examination of the idiom, although it does sometimes apply to the music the tools of analysis and criticism. The book is really a sort of odyssey through almost a century of jazz, our heroes being the musicians who traveled the highways and byways, grew up in myriad hamlets, and resided in the several main centers of jazz activity, all the while absorbing the art form and contributing to its evolution as the decades rolled on. That history, analysis, and critical commentary alluded to above is largely in the words of the artists themselves and, without exception, taken from interviews the author conducted.

The narratives and observations of the artists (and several others) have been organized along traditional lines of jazz history, notwithstanding the procrustean nature of those lines, and have been supplemented by chapters on singers, the overseas scene, and so on. True, there are gaps here and there, and a number of principal players do not herein tell their stories. And yes, there is overlap from one chapter to another, but that is the way of any art form as it evolves from one period to another. The simple explanation for the gaps and omissions is that the accounts were not taken down with a book such as this in mind. They were amassed—some five hundred or so, from which a selection was made—in the course of carrying out the tasks of a working journalist and broadcaster over an extended period, namely from the early 1970s to 1990.

While I often went out of my way to interview certain artists, I never approached my work as a sort of historical jigsaw puzzle, as it were, the anticipated completion of which would constitute a full picture, or history, of jazz. That the stories selected for inclusion here do indeed come together as a continuous, and nearly unbroken, informal history of the idiom was the inadvertent result of the availability for interview in Washington, D.C., during the 1970s and '80s of a very wide spectrum of jazz artists. A mere few of those artists reside there; the great majority of them were visiting the city for a day or two, perhaps a week, performing at one or another of the area's jazz venues.

Turning on the tape recorder, speaking on the phone, and exchanging letters is generally not enough. Catching the artist in performance, gaining familiarity with his or her recorded performance, and researching background in the jazz literature were necessary supplements to the interview.

I recommend to the reader that he or she also seek out the artists in performance at jazz club or in concert hall, check out their recordings, and read up on them in other sources. I have been doing so for a couple of years short of half a century and continue to hear new things, not only from young players making the scene but from older artists playing in their long-established styles. For that matter, recordings I have been listening to since the 1940s still reveal surprises. But enough of the author and his recommendations! Let's get on with the story of this timeless art form, told, for the most part, by people who create it.

Silver Spring   W.R.S.
December 1990

Acknowledgments

I want first of all to express my appreciation to the musicians and others who made themselves available for interviews and who patiently responded to my insistent questions about their lives, careers, art, and opinions. Those responses constitute, in large part, the substance of this book.

The photographers who provided the superb illustrations for this volume must also be thanked. Their art immeasurably enhances this account of the jazz idiom's evolution from its earliest years to the present by bringing to light some of the individuals whose stories and thoughts are herein contained.

It would not be practicable to list here all those others who, over the years, have lent assistance, provided support, and offered encouragement to me in my efforts to gather facts about and gain understanding of the art form of jazz. The list would go on for pages and pages, necessarily comprising not only the hundreds of other musicians I have interviewed, but hundreds more with whom I have spoken informally plus several thousand I have reviewed in performance. In all fairness, a complete roster would also have to include the countless number of devotees of jazz with whom I have discussed this music and whose knowledge I have absorbed. A blanket thanks must suffice to cover the myriad souls alluded to above. What follows are the names, in alphabetical order, of those very special friends and colleagues without whose help and encouragement my task would have been far more formidable indeed. To each of these, *ab imo pectore gratias tibi ago.*

Dick Baker, Bill Barlow, Chips Bayen, Keter Betts, Paul Blair, Fred Bouchard, Clea Bradford, Bill Brower, John Bunyan, Jim Burk, Dave Burns, the late Lou Byers, Charles and Linda Cassell, Tommy Cecil, Spiritus Cheese, Joe Cohen, Willis Conover, Art Cromwell, Stanley Dance, Gail Dixon, Larry Eanet, John Eaton, Elliot Epstein, Don Farwell, Leonard Feather, Ken Franckling, Carlos Gaivar, Ira Gitler, Joe Godfrey, Felix Grant, Harold Gray, Richard Harrington, the late Bill Harris, Dennette Harrod, Rusty Hassan, Bill Hasson, Matt Hessburg, Buck Hill, Patrick Hinely, Shirley Horn, Allen Houser, Mary Jefferson, Willard Jenkins, Mike Joyce, Nat Kinnear, Bill Kirchner, John Kordalewski, Pete Lambros, Jeff Levenson, Dennis John Lewis, Yale Lewis, Jim Lyons, Ann Mabuchi, Frank Malfitano, Bob and Betty Martin, Gene Martin, Bill Mayhugh, the late Johnson (Fat Cat) McRee, Dan Morgenstern, Byron Morris, Dick Mountfort, Steve Novosel, Deater O'Neill, Tim Owens, Frank Panetta, Paul Pearson, Van Perry, Eddie Phyfe, Bill Potts, the late Beale Riddle, Jim Ritter, Marita Rivero, Dave Robinson, Ira Sabin, Bill Schremp, Jack Shulman, Joel Siegel, Rhodes Spedale, Ken Steiner, the late Beale Taylor, Mason (Country) Thomas, Jack Towers, Nap Turner, Fred and Anna Wahler, Jim Walsh, Bill Warrell, Ronnie Wells, Wild Bill Whelan, Andrew White, Michael Wilderman, Hal Willard, Martin Williams, the late James (Mr. Y) Yancey.

Going way, way back in time, I want to add three individuals who provided (and continue to provide) considerable encouragement to me vis-à-vis jazz during my teens when the art form first came to my attention, to wit, my brothers Bill and Turner Stokes and my very first jazz mentor, Richard (Jud) Henderson.

Dot Hawkins deserves special thanks for undertaking and executing the daunting task and tedious chore of typing the final manuscript of my book.

Sheldon Meyer and his staff at Oxford University Press have earned several times over my sincere gratitude for their consistently superior professionalism in carrying out a project which, four years in the making, had often to have tried their individual and collective patience nearly to the breaking point. I especially wish to thank Leona Capeless, Stephanie Sakson-Ford, and Karen Wolny.

Acknowledging that some interviews in this book were originally utilized, in part, in articles or liner notes, I want to thank the following publications and record companies for granting permission to reprint passages that first appeared under their imprints: *Jazz Times, Jersey Jazz, Ms., Soul Note/Black Saint Records, Time-Life Music, Washington-Baltimore Theatre Guide, The Washington Post, Washington Woman.*

Last to be mentioned, but always uppermost in my mind, are three without whose encouragement, understanding, and love this book would never have been conceived, much less commenced, and certainly never completed, my wife, Erika, and sons Sutton and Neale.

Contents

Illustrations follow page 102.

THE JAZZ SCENE

1

New Orleans

My family, they encouraged me to play music because they had a psychological attitude that, if you give a kid an instrument, he becomes attached to that instrument and that becomes a part of him and he'll be respected.
Danny Barker

New Orleans is the only place I know of where you ask a little kid what he wants to be and instead of saying, "I want to be a fireman" or "I want to be a policeman," he says, "I want to be a musician."
Allan Jaffe

The continuing and strong appeal of authentically performed New Orleans style jazz shows no sign of waning. Indeed, the veteran players, who number a mere handful, find themselves as busy as at any time in careers that, for some of them, reach back five, six, and even seven decades. We are speaking of that style of jazz characteristic of musicians who chose to remain in the city of New Orleans rather than join the many who relocated to Chicago and other northerly cities beginning around 1915.

Trumpeters King Oliver, Freddie Keppard, Louis Armstrong, and Paul Mares, trombonist Georg Brunis, reed players Johnny Dodds, Jimmie Noone, and Sidney Bechet, and drummer Baby Dodds had all taken up residence in Chicago by 1922. Trombonist Kid Ory and pianist Jelly Roll Morton spent the late teens and early '20s in California, reaching Chicago in, respectively, 1923 and 1925. Trumpeter Bunk

Johnson's travels commenced at the turn of the century and took him coast to coast and, in bands on ocean liners, to Europe and the Orient.

The dozen musicians cited are the most prominent, in a list that would fill several pages, of those artists who migrated north, west, and east and in so doing came into contact with cultural influences and societal circumstances that forced changes in the music with which they had grown up. (Those changes will be dealt with in later chapters.) While it can hardly be claimed that the musicians who remained in New Orleans enjoyed an isolation from such musical and other influences as their emigrating peers encountered—there were, for example, radio, recordings, and returning musicians to exert their impact—the basic elements, outlines, and overall effect of the early New Orleans style remained relatively stable and still does to this day. The principal elements of the genre are a collective ensemble attack—heterophony, really—a surging yet relaxed rhythm, and an artful blend of carefree abandon generously seasoned with joy and the sweet melancholy that derives from the blues.

Yet one need not resort to analysis to appreciate the New Orleans style—just open the ears to its sound and try to keep the feet still! The explanation for sold-out houses here and abroad which feature, for example, one or another of the several Preservation Hall Jazz Bands (the chief present-day embodiments of the New Orleans style) resides in the riveting impact of the music and the restorative effect it has upon the listener. Its sound is as complete a summary of the human condition as any musical idiom has ever succeeded in creating.

One of the world-traveling bands that plays Preservation Hall in New Orleans when it is not on the road is led by Percy Humphrey, who was born in 1905 and was still fronting the group in 1990. Percy has continued to live in his native city where he has long been considered one of its finest brass band trumpeters. He and his older brother Willie, the band's clarinetist, are the grandsons of James B. (Professor Jim) Humphrey, one of the most distinguished music teachers of New Orleans.

In the 1890s Professor Humphrey, attired in swallowtail coat, would take the train to Magnolia Plantation in Plaquemines Parish to instruct the children of the sugarcane workers in the rudiments of music. "They got to have a little brass band and have a little amusement," Willie told me. The formally trained professor, who died in 1937, saw many of his prize pupils swept up by the new music that came into being around the turn of the century. James Humphrey's students, either at the plantation (where he organized the thirteen-piece Eclipse Brass Band, which sometimes played in the city) or in New Orleans (the brass bands of which he coached and wrote arrangements for), included trumpeters Chris Kelly and Sam Morgan, saxophonist Andrew Morgan and two other Morgan brothers, the trombonists Harrison Barnes and Sunny Henry, his son the clarinetist Willie Eli Humphrey, and three grandsons, Percy, Willie, and trombonist Earl Humphrey, who died in 1971. Many more names could be added, for Professor Jim, who rarely performed in public, was the instructor of virtually an entire generation of New Orleans bandsmen.

"He taught a lot of people, a whole lot of them that made it and didn't want to admit that he was their teacher. He started Kid Ory, he taught Papa Celestin, Kid Howard and Joe Howard, a bunch of them, but they didn't give him credit for it, see," observed Percy, "but he didn't worry about that, he taught so many. All he wanted to do was to help others learn something, help them make a living, that's all. We were brought up in his home, and each of us had to learn how to play at least two instruments. I happened to select drums first and the trumpet or the alto or baritone horn secondly, see. So he started me off with the baritone and I went to the alto and then commenced playing cornet. I finally dropped the drum situation and stuck with the trumpet—I *got stuck* with it, anyhow, and I'm still trying to play the trumpet," he laughed.

Willie recalled that his grandfather "had his system, his way. He would put it: if we didn't do right, well, we could understand it better on our backs!" In his late eighties and still playing regularly at Preservation Hall, even frequently adding his presence on tours, Willie is one of only several remaining New Orleans musicians whose careers stretch back to the teens of the century. He worked on the Mississippi riverboats in bands led by pianist and calliope player Fate Marable, played with Jelly Roll Morton and King Oliver, and knew pioneers of the music who were of the generations of his parents and grandparents. Willie was only seven when the legendary cornet player Buddy Bolden collapsed while playing a parade (and was consequently institutionalized until his death in 1931). "I never met him," said Willie, quickly adding, "but my father did and I did know his mother 'cause she sort of lived in our neighborhood."

Guitarist, banjo player, and singer Danny Barker is nearly a decade younger than Willie Humphrey and was still performing in 1990. I caught him in action at Virginia's Manassas Jazz Festival and pinned him down after a Sunday brunch set for an interview. It was his early years that I especially wanted Danny to talk about, for he left New Orleans in 1930, when he was twenty-one, and his subsequent history therefore plays no part in this chapter.

"I was born in the French Quarter smack dab in the middle of it," Danny began. "My father's people were Baptists, hard-shell Baptists, my mother's people were Catholic, so I was between the two religions. We celebrated the Catholic holidays and I was steeped in the activities of the Baptist church, the screamin' and the shoutin' and the runnin' and the jumpin' and 'Praise to the Lawd!' and meetin' the preachers and everything about the Baptists.

"On the other side, my mother's brothers Paul Barbarin, Louis Barbarin, Lucien Barbarin, Willie Barbarin, they was steeped in music, that was all they talked around the house was music, you know, *jazz*. My grandfather, Isidore Barbarin, was a member of the famous Onward Brass Band, from the period 1890 to 1930 considered the greatest brass band in New Orleans.

"My family, they encouraged me to play music," Danny explained, "because they had a psychological attitude that, if you give a kid an instrument, he becomes attached to that instrument and that becomes a part of him and he'll be respected. 'Little Willie, he play, he's a musician!' Regardless of your status, you play an instrument and people who are non-musical, they marvel at you, they might not tell you, but they marvel at you how you can play something and entertain people. *They* try and pick up an instrument, most time a guitar, which are plentiful, and they get some kind of sound, but they don't have the ability to continue. So they encouraged me. 'Son, what you goin' to play!' Like, 'You goin' to play baseball, you goin' to play football!' Same with music—'What you goin' to play, a saxophone, a trumpet, a clarinet, what you goin' to play, what you want to do?' So that's the atmosphere I came from."

Two of the five major cornetist/bandleaders who remained behind when King Oliver et al. departed the city, Chris Kelly and Buddy Petit, never recorded, and I never let an opportunity go by to hear by proxy, so to speak, what they sounded like. Both died young—Kelly at thirty-seven or so in 1929, Petit at about forty-three in 1931. By all accounts they were the city's finest hornmen throughout the 1920s. In fact, some of Petit's contemporaries thought him the equal of Louis Armstrong.

"Lou Gehrig, Babe Ruth, Joe DiMaggio—that's the way those people were in the neighborhood," Danny proudly exclaimed. "Chris Kelly was a good blues player but he wasn't a great readin' musician, and when you leave New Orleans, they put some music in front of you and they expect you to play it. Chris Kelly had a dozen blues songs and the people he played for, that's all they wanted was some blues, the current low tunes that were around. You learned a dozen tunes and you could play a party. They never advanced, they called it 'routine.' Now in the case of Red Allen"—a trumpeter a year older than Barker and one whose career also took him to New York for associations too numerous to mention—"he played a whole spectrum. You could jump off from New Orleans and go on the riverboat with Fate Marable and go to St. Louis, and you somethin' when you join *that* band, it's like joinin' Toscanini!"

"Buddy Petit was considered by many one of the greatest trumpets that ever existed in New Orleans," Danny continued. "He was called a 'diminished' king. He could run diminished chords which was almost baffling, like Dizzy Gillespie and these modern people, 'cause they extended music. That's how Buddy Petit was recognized, see, by all the greatest musicians, from Louis Armstrong or David Jones, Papa Celestin, King Oliver. Everybody respected Buddy Petit 'cause he mastered diminished chords. They marveled at him.'"

Kid Rena, Punch Miller, and Sam Morgan were the other cornet-playing kings during the 1920s but only Morgan was recorded on location there in the earlier period. This happened in 1927, and it has always intrigued me how and, perhaps more intriguing, why the session took place under the auspices of the Columbia label, which that year was celebrating the 100th anniversary of Beethoven's death with a massive issue of the German composer's symphonies, choral works, and chamber music! Rena, incidentally, was recorded in New Orleans in 1940, and Miller's discography begins with his arrival in Chicago in the late 1920s.

"Well, like fate happens like that," Danny offered. "He [Morgan] probably didn't have the best band in New Orleans at the time to merit making a recording date, but the people met him and he probably acted more businesslike than the others they approached, 'cause a lot of people didn't care about recording. It was, 'Okay, record, what they goin' to do, take my music away and they ain't goin' to give me nothin!' So it wasn't a big thing, and then you have to set up and you have to wait and it wasn't payin' nothin', maybe twelve dollars a man to make a record and they don't give you no royalties or nothin'. Mess over people! And they did that with a lot of country-western people, they done that to them, too. *They'll* tell you what *they* went through with these people." (Many a jazz artist of today would say, "The more things change, the more they stay the same.")

Danny Barker's more than six decades as a professional jazz musician have seen him in the bands of Cab Calloway, Lucky Millinder, Benny Carter, and others. He worked with Jelly Roll Morton, was on the road with Little Brother Montgomery, played dates with Bunk Johnson, appeared in concert with Eubie Blake, played banjo on a 1989 Wynton Marsalis album, and led his own group that featured his wife, the singer Blue Lu Barker. The details of Danny Barker's extraordinary life can be read about in his autobiography, *A Life in Jazz*, a truly classic firsthand account of the art form and the life it entails.

We have so far been looking back to the beginnings of New Orleans jazz. Before we go on to some aspects of its subsequent history, a question begs to be, if not definitively answered, at the very least considered. Were the beginnings of New Orleans jazz the beginnings of jazz itself? I put the question to, among many others, some historians of and authorities on early New Orleans jazz.

"If we think in terms of around 1900 when there were no record players," observed Karl Koenig, whose *Jazz Map of New Orleans* constitutes a guide to more than 800 locations connected with the history of the city's musical culture in the early decades of this century, "when they used music for everything—picnics, socials, political functions, church—it was the right environment and the right time to do live music and it evolved into jazz. It was so powerful I don't think there's any field of music that hasn't been affected by it." The environment that Karl Koenig referred to, of course, was New Orleans.

Preservation Hall founder Allan Jaffe stressed the utility of the music. "New Orleans jazz is a functional music in New Orleans," Allan told me. "This music is still being played in churches there, it's still being used at funerals, at Mardi Gras time, to open up a new store— there's all the clubs that have annual parades, and New Orleans jazz is so suited to all of these needs." In another conversation Allan, discussing the phenomenal circumstances of the Humphrey brothers' survival well into their eighties as active players, alluded to "the opportunity for someone who's been around since the inception of the music" as "rather remarkable." That word *inception*, I think, is the clue to Allan Jaffe's position vis-à-vis the birthplace of jazz.

Never one to let sleeping dogs lie, I also put the question to Professor William Schafer, author of *Brass Bands and New Orleans Jazz*. "That's a question like whether humankind came into being in the Garden of Eden or not," Professor Schafer began. "It's a good myth and it's a sort of nourishing myth, I think, in a lot of ways, and it's one that's very powerful and good to hang onto. I've got nothing against myths— one of my other studies is of myth itself.

"I think that obviously a kind of vernacular black music was springing up all over America after the Civil War and it occurred in a lot of different cities, a lot of different places. The place where we have the most material for study of that movement is New Orleans, because New Orleans has been such a history-conscious city, and, as musicians are always fond of pointing out, the New Orleans musicians were certainly a clannish bunch and they hung on to family and city traditions with great rigidity and remembered and passed on a lot of this information.

"I think that there's enough evidence to show that there were movements in black music all over the country, perhaps none of them quite as consolidated and as powerful as they were in New Orleans. However, I think more of it came together there, perhaps because of the nature of the city. It really was possibly the most cosmopolitan city in the country at that time, I think probably even more so than New York or San Francisco, say, at the turn of the century. And it was a place where the music could flourish and where the musicians were rewarded in all sorts of ways, besides just commercially, for playing. It made a very healthy milieu, I think, for the music.

"But, no, I don't think it started there altogether. I think maybe it held on there better than any place else and grew perhaps faster and in more directions in New Orleans than anywhere else. It's a kind of good story that it started there and went up the river on the riverboats to Chicago and then went over to New York. I mean, that makes a very neat sort of map that you can draw on a blackboard or whatever, but it's not exactly what happened."

The musicians who have been speaking to us up to this point have been some who have, in Allan Jaffe's words, "been around since the inception of the music," or very nearly so. Allan has also pointed out that "in the '20s and '30s there were literally thousands of musicians in New Orleans. Now there's a hundred. You need a replacement and you talk about *that* trumpet player or *that* trombone player. What we need is ten trumpet players, twenty trombone players."

In addition to providing a venue for the older musicians to perform on a regular schedule for a nightly audience young and old, many of them tourists on a first visit to the city, the now nearly three-decade-old institution of Preservation Hall has served as an academy for on-the-job training of musicians from many parts of the world. Some of these—whether from Scandinavia, Great Britain, Japan, Eastern Europe, South America, Canada, or wherever—come for stays of weeks or even months, sitting in whenever the opportunity presents itself. Some settle in New Orleans. Of the several regulars at the hall, one came from California, trombonist Frank Demond, a familiar onstage presence to all who have caught the Humphrey brothers Preservation Hall Jazz Band since 1976 when he took over the trombone chair upon the death of Big Jim Robinson, his mentor. A certain continuity was served by that circumstance, one that is apt to repeat itself as the next few years go by. If the actors in such likely scenarios have been cast in something like the same mold as Frank Demond, then the survival of the style will be assured.

"We heard that this crazy band was at the Beverly Cavern," Frank recalled for me, gazing back into the year 1948 to his teens in his native Los Angeles. It turned out to be the Firehouse Five Plus Two, and Frank and his parents were ushered to ringside seats when they arrived. "The trombone player kept just missing my ear," Frank said with a chuckle. Soon after that ear-opening experience Frank was taken by a friend to "hear the truth—Kid Ory."

Hooked on jazz from the start, Frank took every opportunity to catch the greats passing through the area—trumpeters Louis Armstrong, Muggsy Spanier, Pete Daily, and Lu Watters, trombonist Jack Teagarden, singer Ella Fitzgerald, and "an awful lot of bands."

"I got interested in the George Lewis band on record first and then they came in the early '50s and I would go and listen to them every night." Frank had begun to learn banjo soon after that first experience of hearing a jazz band at the Cavern and had within a year organized his own band. "When Lawrence Marrero took sick, George asked me to play banjo with his band for the last two weeks at the Beverly Cavern. That was really an exciting time. I was maybe twenty years old and it was an initiation into the real hard-driving New Or-

leans style. I went through a stage of liking the King Oliver band and
Jelly Roll Morton and all that and I still do, but that rough-hewn
spontaneous New Orleans sound has always just really grabbed me
and it's the kind of music I really enjoy the most, although I like some
of the other styles too. That's what's in my heart, that's what I really
like to do. I like the New Orleans swing, the beat."

Frank made the shift to trombone in 1951, and at Pomona College
he founded the Storyville Stompers. An album they made in 1955 on
the Tropicana label is a rare collector's item. Out of college, Frank
was regularly playing in bands made up of visiting New Orleans musicians, and in the early '60s he formed the Salutation Tuxedo Band.
In 1965 he visited New Orleans and sat in at Preservation Hall, in the
process becoming re-acquainted with clarinetist George Lewis. In 1968
Frank took a year off and brought his family to New Orleans. Mind
you, for nearly fifteen years this musician had supported his family as
a designer of ocean-front houses.

"There again, I was self-taught," Frank points out. "It seemed that
everything I ended up doing, it was just from an intuitive way—I'd
just watch and do it." He studied economics in college, not architec-
ture, and has never taken a musical lesson in his life, yet has played,
in addition to trombone, banjo, guitar, and piano.

"When Allan Jaffe asked me to play in 1971 on a tour with Billie
and Dee Dee Pierce I had three custom houses going," confessed Frank.
"Obviously, my heart was with music because just on one day's notice
I said, 'Sure, I'll meet you in El Paso,' and phoned up the people and
said, 'Well, there's been a little change.' I eventually finished those
homes, but I never got back into that business." In 1974 Frank moved
to New Orleans, and in 1976, he became a regular member of one of
the three Preservation Hall Jazz Bands.

"There's a certain heart that this music has that it doesn't seem to
matter whether anybody knows anything about jazz or not," Frank
mused, "it just reaches people. It doesn't matter if we're playing in
England or Ireland or Caracas, Venezuela—the people just respond.
We played in Managua and the people just couldn't believe it—they
thought they hated Americans and they loved the band. It's on an
intuitive level. If you listen just with your intellect, you're going to
miss it. It has an inner beauty that touches an inner place in people.
That's the reason I play it. It's just playing in a 'love' band. It's like a
celebration and it's really wonderful to be in that every night."

There are, of course, others, known and unknown, whose backgrounds approximate the sort of intimacy with the New Orleans style
that has marked Frank Demond's history. As we suggested above, some
of these are in New Orleans waiting in the wings. Yet let us not over-
look what is happening in the city among its natives.

"There was a time when I was concerned whether the music would
continue," Allan Jaffe admitted to me, "but now I'm sure that it will.

There was sort of a 'skip' generation, that generation that was coming up in the '40s, that wanted to be bebop musicians or rock musicians, but the younger generation now wants to play this music. First of all, they can make a living doing it, and they realize that it's still an exciting and a vital part of the music scene. I'd say that during the past ten years there's been an awakening in the black community that people who want to be professional musicians can continue playing New Orleans jazz as their fathers and grandfathers did. Just recently Willie Humphrey was playing at the hall one night with the young trumpet player Wendell Brunious and Willie said to him, 'Hey, did you learn that phrase from your father?' and Wendell said, 'Yeah,' and Willie said, 'I taught it to him.' So here's a musician in his late twenties and Willie has sixty more years of experience than him. Willie has played with just about all the great New Orleans musicians. He played with Ma Rainey and he played with all the people in Buddy Bolden's band except Bolden."

Allan was encouraged that "there are bands in New Orleans who are all teenagers and are trying to get back and play like the old brass bands. This is the first time this has happened since the end of the '30s. And, hey, the music really sounds great. As long as I can listen to it and it sounds this good, I'm very happy. I hope the young people continue playing it, and I think the thing that has been an encouragement to the young musicians is seeing the acceptance again of the music."

Now Allan was anything but rigid in his outlook, for he accepted and even welcomed the potential for change in the style. It may surprise, even shock, some of the diehard antiquarians out there that Allan Jaffe was a fan of the Dirty Dozen Brass Band, who superimpose decades of musical evolution upon the traditional New Orleans brass band format, extending the repertoire through big band, r&b, bebop, and free form.

"Here's someone who's taken the instrumentation and the tradition that already exists and come up with a totally new approach still serving that tradition. They play for funerals, they play for parades, same as the Olympia or the Eureka brass bands. It's not the music of a *period*, it's really the music of a *city*, and it's as current as the day it's played. It's modern, contemporary music because it has a modern, contemporary use. And as long as the music has a contemporary use and can be functional and effective, I think it's going to continue." Allan paused and then appended to his remarks an observation that sticks with me still. "New Orleans is the only place I know of where you ask a little kid what he wants to be and instead of saying, 'I want to be a fireman' or 'I want to be a policeman,' he says, 'I want to be a musician.' "

The man who said that to me a couple of years ago was a major mover and shaker in restoring the vitality of the traditional jazz scene in his adopted city. Allan Jaffe died in March of 1987 of cancer at the age of fifty-one. Allan's helicon (the precursor to the sousaphone) has been replaced in the Humphrey band by the upright bass of James Prevost.

Another who has done more than his share in reviving the tradition and aiding its perpetuation is Danny Barker. When we finished our discussion about his past I asked him to talk about what he had been up to these past several years, because I had been reading about his good works there in his city. Danny was not loath to expatiate upon his role in encouraging the young of New Orleans to become involved in music. After all, had he not gotten his start through the prompting of his elders?

"The minister of the church—John the Baptist Church, 'cause I like the spirit of the church— . . . said, 'Mr. Barker, I understand you're a musician.' I say, 'Yeah.' He say, 'Well, you think it's possible you could form a marchin' band for the church for when we have our affairs, and get the kids in the neighborhood interested? I hear a lot of people playin' horns around here but I don't see 'em together. My father had a brass band and we used to play all the Baptist affairs.' I say, 'I think I could.'

"So there was a little kid in the neighborhood named Leroy Brown, he used to play his horn and all the other little kids come to look at him and he played well, he had a good tone. He went to Catholic school and the teachers was interested in him and he could read real great. Another one had drums and they used to play around there but they had no guidance. Leroy hadn't the least idea how to form a marchin' band.

"I didn't know Leroy and when they asked me to form a band for the church I went up to him and asked him his name. I said, 'Leroy, do you think you'd be interested in formin' a brass band, a marchin' band like the Olympia, the Imperial?' 'Gee, yeah,' he say, 'I'd like to.' 'Well,' I say, 'will you get two trombones, two clarinets, and a tuba? I got a brass drum and a snare drum player.' He say, 'Sure' and inside of a week he had 'em. My two little cousins, Charlie Barbarin and Lucien Barbarin, one played the snare drum, one played the bass drum, they came and joined the band, and in two weeks they had a little marchin' band. So they're marchin' the neighborhood every evenin' and the people were thinkin' it was a big marchin' band, people get excited, 'Oh, what's happenin'?' I love a parade, you know, that sort of attitude. So all the neighborhood, all the little kids, I mean, they was second-linin', people knew where the kids was—around Leroy, about thirty or forty of 'em. It's a free show, somethin' new for the kids.

"Leroy's in Hong Kong now, he's a big favorite over there with a jazz band. There's about thirty of them back here who are playing professional music now, they're workin,' they're goin' to school, and they got good jobs 'cause they're educated now. It ain't like when I was comin' up—they go to high school, they finish school. So that's what's happenin' now, there's ten kid bands in New Orleans—the All Stars, the Pin Stripes. But the city's not the least interested. They'll give the symphony a hundred thousand, they'll give to the classical music, the ballet, but they do nothin' for jazz, which is tragic.

"But the roots is set now, its roots is set. The little All Stars band is four brothers, three cousins. They've been on the Norway, that big ship, they've been to Europe, they've been to Japan. Those kids, they tend to business. The oldest one's James Jr. and his little brothers and cousins, when he says somethin', he means it, you know, and they understand. And they got money, the kids got a pocketful of money. They go in the French Quarter, they got a big bucket, people come there and pour money in the bucket, in two hours they're loaded. So it has been revived. The city should pick up on it, but I doubt it, they want somethin' for nothin', they want the people to play for nothin'."

The irony in this circumstance, that the—for many—"birthplace of jazz" has declined to lend support of a tangible kind to nurture a rebirth of the city's own musical style does not go unnoticed. That the music of New Orleans and its offshoots will survive there and elsewhere we do not doubt. In fact, let us cite a few random examples of the music's tenacity and its potential for growth. For the music of New Orleans has proved, over the decades, to have an amazing capacity for evolution, sometimes in other styles, such as r&b, sometimes in individuals.

One of the most astonishing musicians to come out of New Orleans in recent years is the pianist Henry Butler. It may seem all the more astonishing that Henry suddenly appeared, seemingly out of nowhere, upon the national scene in the mid-1980s and has been accorded much exposure over the airways and has garnered a great deal of attention. Actually, jazz players of note have a way of doing just that. Jack Teagarden's arrival from Texas in New York in the 1920s, Cannonball Adderley's from Florida in the 1950s, and Butler's fellow New Orleanians Wynton and Branford Marsalis, Terence Blanchard, and Donald Harrison come readily to mind. All four made their national debuts in the 1980s in Art Blakey's Jazz Messengers.

Henry Butler was taking piano lessons at the Louisiana State School for the Blind in Baton Rouge when he was eight; at nine he added drums, and at twelve the baritone horn. At Southern University he majored in voice. His influences range across the spectrum of Afro-American music from the blues to Scott Joplin to gospel to r&b to Thelonious Monk, Miles Davis, and John Coltrane, as well as to Bach and other European masters. Henry came directly under the influence of New Orleans pianist Professor Longhair, and this puts him "smack dab in the middle" of the New Orleans tradition. For Professor Longhair took lessons from Tuts Washington, who played in the bands of Kid Rena, Isaiah Morgan, Punch Miller, Papa Celestin, and others in the 1920s. The forty-year-old Butler will likely never be found playing in a New Orleans style jazz band (although he could do so with one hand tied behind his back), but the flavor of the New Orleans idiom will always be detectible in his playing—of that I am certain. "I feel that I'm free to do anything," Henry told me, "and it's not because I'm confused, it's because I want to be omni-directional."

Another New Orleans pianist, half Butler's age, who has recently emerged from the city's embrace is Harry Connick, Jr. A child prodigy, Connick learned to play his first tune, "When the Saints Go Marching In," when he was about three. At nine his father recruited some New Orleans traditional-style musicians, including drummer Freddie Kohlman, bassist Placide Adams, and Pee Wee Spitlera, all veteran players, and they cut an album on a local label. Harry was playing nights in Bourbon Street clubs throughout high school, and he went on to study under Ellis Marsalis and to study theory and voice at a public high school for the arts. Again, you may not find Harry Connick spending much time in a traditional New Orleans format. His debut album in 1988 in a trio setting indicates that his roots, like those of Butler, will continue to show, but he has already become a modernist who owes as much to Monk as he does to Jelly Roll.

Then there are the Marsalis brothers, Wynton and Branford, Donald Harrison, and Terence Blanchard, and so many other young players from New Orleans lending their artistry to the evolving continuum of the music. Continuum is a word, and a concept, to bear in mind when listening to jazz of any style. It's all connected, from the raucous gut-bucket blues of Buddy Bolden's band in the 1890s (the oral tradition provides that description—Bolden never recorded) to those far-out licks the cats were trying out down at Joe's Bar last night. If you don't believe me, tune your ears into some old-fashioned New Orleans style, then hang out at Joe's again tonight.

Eric Dolphy, a multi-woodwind-playing creative genius whose associations included Charles Mingus and Ornette Coleman and who died in 1964 at thirty-six, is reported to have come away astonished from his first experience of a New Orleans marching band (of elder musicians, mostly), observing to his companions that these were clearly "the first free players," a distinction that he had before that occasion always thought belonged to his peers and to himself.

Another visitor to the city, in the mid-1970s, the New Orleans style clarinetist and saxophonist Tito Martino, who was at that time leader of the Traditional Jazz Band of Sao Paulo, Brazil, put it differently.

"We have discovered in the early jazz a particular sense of balance between musical form and feeling, between what we call 'soul' and musical elements," Tito explained to me during the course of a 1975 month-long tour of the eastern U.S. that I arranged for the band.

(The experience, an eye-opening one for me, constituted my only participation in the entrepreneurial aspects of the music.)

"So we try to stick to this tradition and we feel that . . . this . . . does not in any manner make the music outdated. We feel that we are playing 'today' jazz with a grammatic form that was invented by the black people of the south of Louisiana, of New Orleans, in the beginning of this century. So is music of today made in the manner of yesterday."

2

Chicago

The music was all over the place. It's just unbelievable how many bands there were in Chicago.

Wild Bill Davison

The people that really turned me around in my early days were nameless. I wouldn't say they were fine musicians or studied—they were blues musicians.

Art Hodes

Before Louis passed he said, "We didn't know we were making history, did we?" And I said, "We certainly didn't!"

Earl Hines

Beginning about 1910, New Orleans jazz musicians began arriving in Chicago, some to stay, others only to visit. Among the first were the pianists Tony Jackson and Jelly Roll Morton. Jackson remained until his death in 1921. The itinerant Morton lived in the city for several years in the early teens, moved to California in 1915, used Los Angeles as his base until settling in Chicago in 1923, and relocated in New York in 1928.

Another early visitor to the city was the trumpeter Freddie Keppard, who had taken a New Orleans band onto the vaudeville circuit in 1912 and settled in Chicago in the late teens. By 1918 trumpeters Sugar Johnny Smith, Manuel Perez, and Tommy Ladnier, trombonists George Filhe and Roy Palmer, reed players Sidney Bechet, Alphonse Picou, Lorenzo Tio, Jr., and Jimmie Noone, bassist Bill Johnson, and drummer Paul Barbarin had come from New Orleans to grace various Chicago cabarets or theaters with their artistry. King Oliver arrived in 1919 and by 1922 had sent for clarinetist Johnny Dodds and his brother, the drummer Baby Dodds, and Louis Armstrong to join his band. New Orleans trombonist Kid Ory, who had taken up residence in Los Angeles in 1919, moved to Chicago in 1925 and remained for five years until returning to California.

All of those mentioned so far were black, and the dozen and a half listed are a mere fraction of the New Orleans jazz musicians who performed in Chicago during the early years of the music's history in that city. As we pointed out, some were there merely on visits (Perez, Picou, Tio), and some died young (Smith in 1918 in his late thirties). For others, Chicago became home for a decade and more.

A smaller number of white New Orleans jazz players also made it to Chicago, and the first of these was apparently trombonist Tom Brown, who brought a band which included drummer Ragababy Stephens. That was in 1915. The next year saw the arrival from New Orleans of the musicians who would put together the Original Dixieland Jazz Band, which in 1917 would make the very first jazz recordings. They included trumpeter Nick LaRocca, clarinetist Alcide Nunez, and drummer Tony Sbarbaro. In 1919 trumpeter Paul Mares and trombonist Georg Brunis left New Orleans for Chicago, and before two more years were up they were joined by their childhood friend the clarinetist Leon Roppolo, the three of them in 1921 forming the nucleus of the New Orleans Rhythm Kings, a band that would exert considerable influence upon the young white Chicagoans taking up jazz in the 1920s.

Now while we have included in our summary account of the Chicago jazz scene of the teens and early '20s the names of less than thirty musicians, most of these were major players in that scene and many of them went on to decades-long careers and international fame. And may we again point out that the major source of this astonishing supply of talent was New Orleans. Yet we hasten to add that musicians came from elsewhere to play important roles in the great melting pot of Chicago jazz in the '20s, and there were, as well, musicians who grew up in the city or its environs who mingled with and, in some cases, became all but inseparable from their soul brothers from the South.

We are going to let three such artists tell their stories. They are: Wild Bill Davison from Defiance, Ohio, Earl (Fatha) Hines, from Pittsburgh, Pennsylvania, and Art Hodes, who was born in Nikolayev, Russia, and was brought to Chicago by his parents when he was about six months old. Their paths crossed in those early years and continued to do so for five and six decades as they traversed the globe.

"I came from a little town," Davison began one of several interviews over the course of which he told me his life story. "Of course, like all kids, we had a school band. In those days"—Bill was born in 1906—

"the schools didn't sponsor bands like they do now and the most troublesome part of it was that there was no such thing as music teachers in my home town. So we all had to teach ourselves more or less.

"In my home town I think the first thing that we heard, the young musicians that I was hanging out with in those days, would be Ted Lewis, and of course that didn't satisfy us, that wasn't exactly what we wanted to play. But then there was a lot of records coming out by the Original Dixieland Jazz Band and I think the first one, the first record I ever had, was 'Barnyard Blues,' We'd go down to the Victor shop where they sold the records and get their catalogue and then we'd just wait 'til we got the next one, see what they were doing.

"One of the first little bands we got together was called the Ohio Lucky Seven, which was way back in the early '20s. We played all through Ohio, Indiana, and Wisconsin, the Middle West. I never got really started until I went to Cincinnati where I played with the Chubb-Steinberg band, and that's where I met Bix [Beiderbecke]. That was in 1924 and I recorded for Gennett at that time. The band was sort of an arrangement band, really, but there was a few good swing players in the band.

"From Cincinnati I went on to Chicago. There was more jazz bands in the country at that time. Every town and every little hole in the wall had a band in it. I remember talking to a union man who said that there were a thousand jazz bands working in Chicago. Now that doesn't mean big bands. I mean, you could walk in a fairly good-sized bar and there'd be three musicians playing in there. The music was all over the place. And there were hundreds and hundreds of ballrooms. The big bands had a heyday. The Charleston came along and the Black Bottom and the Bunny Hug and all those crazy dances they used to do. There were hundreds of musicians coming in and out of Chicago because Chicago had the biggest ballrooms in the U.S., like the Trianon and the Aragon, where'd you'd look out and see two thousand people on the floor. And all the nightclubs had good-sized orchestras and everybody in those days was very dance-conscious. They not only had a show band but then they'd have a big band for dancing and a small band for rhumbas and tangos and things like that. And there'd be fifteen, twenty girls in the line and ten principals, just like a big Broadway show.

"The music was all over the place. It's just unbelievable how many bands there were in Chicago. You could just go *anywhere* and there'd be a band. I used to go down to the Sunset and listen to Louis [Armstrong] almost every night. Joe Oliver—many times I used to hear him at the Plantation, which was right down the street from where Louis was playing, and I used to go down to the Grand Terrace and listen to Earl Hines.

"Then there were all the blues singers and all the blues bands. I used to know Pinetop Smith and go down and listen to him play at the rib joint. That was a funny place. You'd see people all dressed up there, but there was no floor, it was just a dirt floor, and they had his piano sitting on a little wooden platform. And you'd actually see what they called the 'carriage trade.' You'd see *many* well-dressed people come there to listen to him play.

"In those days there were many jam sessions. A lot of clubs that couldn't afford a big band would have maybe a rhythm section, and it was a good place for musicians to meet and play a little, play with their friends that they didn't work with ordinarily. There were different classes of clubs, joints where there was just cheap whiskey, and cheap speakeasies. One thing that no one has mentioned much . . . is those places they called a 'beer flat.' This would be a big apartment rented by someone and they would take all the furniture out and just put tables and chairs in there. They always had maybe a good guitar player and a clarinet player so they wouldn't make too much noise. It was a wonderful place to go and drink. You'd get a bucket of beer for thirty-five cents and a drink of scotch for thirty-five cents.

"No race problems. You were as safe on the South Side as you were in your mother's lap. They were tickled to death to have white people come down there because that's where they were getting their income from and you were treated with a great amount of respect.

"The Capone mob had something to do with some of those clubs. The big one was the Midnight Frolics at 22 South, a very famous one, and I played there many times. That was sort of a hangout for Capone, really. Oh, I was in places that not only got raided, but a gunfight would get started. Once or twice I saw a killing in a club and some terrible things like that. Like a gangster come in one night before a show and started shooting the lights out, things like that. If you don't think that would empty a place out in a hurry! But I never knew any musician that ever got hurt or in trouble, if he just kept his mouth shut and tended to his own business.

"At first I played some jazz in Chicago, then I got into the theaters, when they had the stage show days. The reason for me not staying in jazz per se is that the theater paid so much. The clubs didn't pay that kind of money, unless you happened to be in a club that tipped well. Like I once played in a place with a good jazz band and we didn't even consider our pay, we only considered the amount of tips per week which per man amounted to about two hundred dollars. That was a lot of money in those days. The guys that were working in jazz around Chicago in those days weren't making very good money and the work was spotty. I saw a chance to really make some real money, so I went with a theater orchestra. Imagine me, eighteen, nineteen years old, making eleven thousand five hundred dollars as a sideman. That's more than senators made in those days.

"We would play an overture in the pit and then after the organs and the bouncing ball thing we would go on up on the stage and do a regular production, dancing girls and plenty of principals and a big band, thirty men. We played almost the same kind of music that a big orchestra would play in the pit of a theater for a Broadway show, only we did that on the stage. The shows had musical numbers that featured just the orchestra. I played all the jazz for the girl dancers. They'd always call on me and the rhythm section to play behind the dancers.

"I was one of the lucky ones because our band stayed together until about '31 in the theater. Most of the theaters closed but the Benny Meroff band that I was with was considered one of the finest theater bands in Chicago. They kept a couple bands going because all the theaters didn't close up. Of course, what really closed them up was the talking pictures. When that came in that really put all the bands out of work. But I went almost completely through the Depression without knowing there was one. I played the Oriental and the Chicago and the Uptown and the Paradise, the Granada—all those theaters had big orchestras.

"Just a couple of weeks ago [in 1974] I went by one of the old theaters that I played and I thought, 'Gee, it'd be fun to get back inside that theater just to look around.' It so happened that it was closed, but we went out in front just to look at the place and there was an old colored fellow inside washing the windows and we asked him to come out. And he said—now, mind you, this was from 1928 that I had been in that theater—he said, 'I remember you. You're the guy that used to go listen to Louis Armstrong every night.' How he remembered me I don't know.

"I used to go out every night soon as the theater was closed. I'd either go hear Louis or go some place and play the rest of the night for nothing, just in order to play. I'd go jamming or go listen to somebody, because I could afford to do it and I enjoyed it very much. Anyone who lived in Chicago at that time was just damned lucky to be able to hear all the stuff that was going on.

"In the early days, of course, I met Bix when he was with the Wolverines playing at the Stockton Club in Cincinnati and I was *very* young at that time, I think around fifteen, sixteen. And we became very good friends because he was a friend of the banjo player in the band I was with. I had just started to play trumpet and mellophone and banjo. So I got to play with him quite a lot. I always thought that he was way ahead of himself, years ahead of any trumpet player in those days. There was a funny thing about the trumpet players in those early days. They all thought that jazz must be played with a mute in the horn. Of course when I heard Bix for the first time that's when I threw my mutes away forever. I've never been a guy to use mutes. So my influences—the first two trumpets that really excited me at all— was Louis and Bix. I liked Louis for what *he* did—I liked that driving style and I've kept that going—and I like Bix for all his pretty notes.

"Of course I got to know Louis and Zutty [Singleton] immediately and I would hang out down there in that part of town, and I sat in with Louis at the old Sunset Cafe. I sat in with almost everybody, any place that I could, because in the theater I didn't get to play much jazz.

"A lot of the early colored bands around Chicago were marvelous. Jimmie Noone, Johnny Dodds, and those guys. I was with the Seattle Harmony Kings, which was a helluva band, in 1927 at the Rendez-vous, and we used to alternate with the Ben Pollack band with Benny Goodman playing the clarinet. We might be there for ten weeks and then they'd come in for ten weeks. That Pollack band was just unbe-lievable for that time. They were so far ahead of themselves. Bud Freeman and Fud Livingston, the famous arranger, was in that band and Glenn Miller was a trombone in the band, a helluva band, just marvelous.

"I think I met Eddie Condon about 1925. He was living in a place called Chicago Heights, a little tiny town out near Gary, Indiana. I never worked with Eddie around Chicago. And I didn't meet Pee Wee [Russell] back in the early days, but I knew of him, of course. I met him in 1941 when I joined him at Nick's in the Village [in New York]. There's so much comment about whether Pee Wee was a good player or a bad player. I think there was no clarinet player that ever lived who knew how to play a last chorus or an ensemble like he did. He always could find the right note. I wasn't nutty about his choruses because I always thought sometimes he would never finish what he got started, but he always did somehow."

Wild Bill Davison left Benny Meroff in the fall of 1931 and put together a big band of his own. Frank Teschemacher was lead alto saxophonist. The band spent some considerable time rehearsing and got a few gigs. Hal Willard, who has been doing research since the mid-1970s for a Davison biography and has interviewed more than three hundred persons who came into contact with the cornetist, filled in some of the details for me regarding Bill's last year or so in Chi-cago.

"The big band was good, according to Bill and according to the only other survivor of that band, trombonist Mort Croy," Hal told me. "Bill and Tesch were driving home in a convertible fairly late after dinner at someone's house. They had had drinks but weren't drunk. Bill, driving, came to an intersection and a cab without lights that Bill didn't see hit them broadside, knocking them across the road and into a tree. Bill flew out of the car and through a store's plate-glass window. Tesch, who was riding slumped in his seat with his hands in his coat pockets, was thrown out and landed on his head against a curb.

"Both were taken to a hospital and Bill, only dazed and bruised, was released very soon. Tesch died about three hours later. Bill was taken to the police station, made a statement, and was released. He heard there that Tesch had died and made the famous quote: 'What are we going to do for a lead sax player?' Or words very much like that. His feeling and expression was one of horror and I think totally within the context of his life. To Bill, life is playing music and nothing else matters.

"As near as I have been able to determine, no musician heard Bill make the remark. I don't know who, how, or where it first was published or passed around. Bill blames Muggsy Spanier for putting the worst possible meaning on the quote. Mort Croy and other members of the band waited at the rehearsal hall for Bill after they heard about the accident. Mort remembers Bill coming in and telling what happened and being very broken up about it. Toasty Paul (Kensel C. Paul) replaced Tesch as lead alto, but it wasn't the same, and when Bill was suspended from the union (for taking Class B money for a Class A job) the band came to an end.

"Bill says everyone remembers Tesch as a clarinet player but that he really was the best lead alto player he ever heard. I also have interviewed a guy who encountered Tesch in Wisconsin playing a magnificent tenor sax.

"There was a coroner's inquest after the accident and Bill was found totally innocent of any wrongdoing. Chicago police records of the era have been destroyed. At Tesch's funeral Tesch's widow and mother approached Bill and told him that they did not blame him in any way.

"But some musicians faulted Bill in this way: 'You know Davison, how reckless he is and how he drinks.' Still, Bill was not ostracized in any way I can find out about, and far from exiling him to Milwaukee, he stayed in Chicago jobbing around wherever he could until October 1933, when a guy named Carl Dunlap asked him to come to Milwaukee to lead a band he was putting into the Wisconsin Roof Ballroom. One of Bill's activities during the period of March 1932 to September 1933 was as leader of the Beau Brummels, a four-man strolling group that played the Sherman Hotel and the World's Fair (Century of Progress) in Chicago. The fair job was in the summer of 1933, a year after the Tesch accident.

"Most of the 'name' musicians who allegedly ostracized Bill weren't even in Chicago when the Tesch accident occurred. They were in New York. They could have heard bad words about Bill through the grapevine, but there is no evidence of them ostracizing him. When he finally went to New York in 1941 he not only wasn't ostracized, his first job was leader of the band at Nick's. His reputation as a hot trumpeter preceded him and when he walked into Nick's and was standing at the bar, Nick came up and said he wanted Bill to sit in, and afterwards offered him the job as leader starting the next night."

Wild Bill Davison was one of those larger-than-life types whom novelists base a character upon and then puff up until they are hardly credible. The problem with that analogy, however, is that in this case the model was already beyond belief. On the bandstand or off, blowing his cornet or playing the raconteur, the "Wild One" was loud, brassy, and not a little vulgar. His humor (vaudeville hokum was part of his routine in the early days) ran to the off-color, breached good taste, or employed insult.

Seated in the crowded dining room of the Maryland Inn, Annapolis, in the early 1980s, Davison related fortissimo an infamous prank of Joe Venuti that involved Gene Autry's sexually aroused horse (who was about to accompany his master onto the stage) and the violinist's teasing bow. Back on the bandstand Bill referred to a recent trip to Argentina where he sought out items for his collection of German helmets and other WWII memorabilia. "I wanted to find a Nazi general and have him stuffed," Bill quipped. When the laughter had died down he observed, "That's my wife laughing—I could recognize her laugh across a football field." To a fan voicing a request he barked, "Get your own band."

There used to be the drinking jokes, the not-very-subtle solicitations for a round for the band, his delighted announcement that 'All the drunks show up to hear me,' and his open admission that 'I'm a drunk, too.' But Bill gave up drinking at the age of seventy-seven. "Quitting nearly killed me," he said, but conceded that it was the only thing to do since his legendary robust constitution was finally informing him that it had had enough.

If so much of this seemed more than a little outrageous, so also was Bill's jagged, explosive, raw, and utterly charming instrumental voice, a one-of-a-kind and inimitable sound that was, thankfully, still with us in the 1980s in the flesh, albeit not frequently in performance here in the States. Davison was still blowing strong in the late 1980s, spending a good part of the year on the world club, concert, and festival circuit. A working musician since his mid-teens, Wild Bill Davison by then had nearly seven decades behind him as a professional jazz player, a record that few can match. Wild Bill Davison died in November 1989, a month and a half short of his eighty-fourth birthday.

Another long-lived player, a "survivor," to use his own term, an artist of great distinction, and a Chicago legend, pianist Art Hodes was also still quite active in his mid-eighties in 1990. As has been the case with some other veteran jazz artists, Art's career had blossomed in that decade after the slim (for jazz players) '60s and much of the '70s.

Russian-born, Art was brought to this country around 1904 at the age of about six months by a father who "didn't like the society there" and a mother who "had a dream that I would become a great pianist." A tinsmith by trade, Art's father played Caruso on a wind-up victrola and sang fragments of operatic arias around the house. His mother was also a music lover and an older sister played popular tunes on the piano.

Art came up poor but proud in Chicago's turbulent 20th Ward where "most of the guys who were dealing in illegal booze grew up. In fact, one of the kids I ran around with I later saw sitting at a ganglord's table. It was a job trying to go to school without getting hit over the head with a sandbag."

Art did not really get into music until his early teens when he started picking up broadcast remotes of the Coon-Sanders Orchestra on the family crystal set. He was soon taking piano lessons at Hull House for twenty-five cents a session. One day a ten-year-old Benny Goodman walked into Hull House and asked if he could sit in. The two jammed for half an hour on some pop song of the day. "As small as he was, it seemed like he was looking down on you," Art wryly observed.

By his mid-teens Art was accompanying singers who "made the rounds, going from table to table for tips," playing dime-a-dance halls, and generally "beginning to get exposed to musicians who may have heard jazz. I'm playin' with a clarinet player from New Orleans that's up here. I'm playin' with different drummers, saxophone players. I'm gettin' it second-hand. And I'm swingin' with these people."

The artistic leanings of his parents notwithstanding, Art's father insisted that his son attend a technical high school. "I had about as much right to be there as a fly in my soup." He began to cut classes and take a tram, spending his afternoons in burlesque houses where he would sit in the front row and dig the three- or four-piece pit band playing all manner of acts that sometimes included pianist and comedian Jimmy Durante. Pianist Joe Sullivan, a native Chicagoan about the same age as Art, was in some of those pit combos, as were some of the New Orleans musicians mentioned at the outset of this chapter.

But for the most part, Art clarified, "the people that really turned me around in my early days were nameless. Sure, I heard Pinetop Smith and the other boogie-woogie pianists, but I'm talking about people I would hear night after night. They were poor and they would come into the barbecue place that was sort of a central hangout for itinerant musicians. One would come in and then you wouldn't see him for a while and you'd see another one. Whatever they made was mostly tips. I wouldn't say they were fine musicians or studied—they were blues musicians."

A big turning point came in the mid-1920s when Art took a summer resort job with banjoist Earl Murphy at Delavan Lake, Wisconsin. Murphy packed along a small collection of 78 rpm records and a wind-up machine to play them on. Represented were King Oliver, Louis Armstrong, Bessie Smith, Earl Hines, Jimmie Noone, James P. Johnson, and Ethel Waters.

"It seems I was black-oriented from the beginning. I came back to

Chicago and met Wingy Manone," Art recalled of his initiation, through the white New Orleans trumpeter Manone, to the black jazz scene on Chicago's South Side. "That was the beginning of my meeting black musicians." For Art this was a pivotal period in his musical development; he thinks of himself as consciously wanting to change his direction. For while he had been getting the music by "osmosis" on the streets and in the dives and by proxy from musicians who themselves had come into contact with New Orleans players, he had heard only a few of these artists first-hand. Suddenly he was experiencing total immersion in the idiom that would become his life. As Art tells it, life with Wingy was "a gas."

"Living with Wingy, whoever gets up first puts the 78 on the turntable and winds it up. And that went on all the time we're in the room. When we left the room it was usually to go out to the South Side and be with Louis Armstrong, at least four days a week. Our whole life was hearing it and going out to be with him and then going somewhere to play. So this was what turned me around and this went on for two years. For two years I never read a book, I never saw the jokes in the newspaper. Our time was completely taken up with listening and playing.

"The next step was that Louis took us to a rib joint and the minute I hit that rib joint I was home free because they had a player piano and you put your nickel in and it came out black. And they had a juke box and the music was black. And it seemed like the word would go out if there was any trade came in and a trio would appear from nowhere. Jackson—the only name I ever knew him by—was the piano player, and he had the blues. This guy was down on his ass."

Before long Art had met Joe Oliver, Zutty Singleton, Earl Hines, and many others. Memories loomed for him. A jam session at the Liberty Inn when four young white trumpet players got up on the bandstand together—Paul Mares, Louis Prima, and Manone, all from New Orleans, and Chicagoan Marty Marsala. Then pianist Bob Zurke came in and "played so much boogie-woogie I couldn't stand it and had to run out," Art chuckled at the recollection. On another occasion, sitting at the feet of Fatha Hines in a club, "Some gangster recognized me and pulled me by the arm up to the piano, says, 'Earl he's gotta sit in.' "

As the Roaring Twenties went by, Art found himself in the company of, and often playing with, Beiderbecke, the Dodds brothers, Noone, Omer Simeon, Cow Cow Davenport, Meade Lux Lewis, Tut Soper, Gene Krupa, George Wettling, Big Bill Broonzy, Condon, Davison, Russell, Spanier, Freeman, Teschemacher, and many, many others.

"Chicago was growing-up time," Art explained. "I played everything from the Bucket of Blood to the Chez Paree. There was no jazz work. When you had jazz work it was because you worked with a group that played jazz but they played the book that the club needed. In other words, if you're playin' dance music, you're playin' dance music. If you played jazz, you *put* jazz *in* the music. Or late at night, at certain clubs, you could get hot."

Art could tell many stories about Capone and his thugs, and he readily admitted that "there'd have been no work for me if it hadn't been for those hood-type people who liked this music and hired us." But his comments remained general: "I saw them, I saw their antics, I saw where a kid I liked very much hung around with them and then put in his application to become a policeman and later they found his body. I saw things like that in the paper. But I actually saw nothing. You observed and you kept your mouth shut and minded your business. And they left you alone, if you did that, because you were an employee."

In 1938, a decade after most of his fellow musicians had departed the city, Art left Chicago for New York, where he remained for twelve years. Chicago had become "a ghost town, music-wise," he avers, "but I'm still hangin' around because I'm still gettin' my education, still hangin' out at the South Side, still listenin' to the blues players, still chasin' it down. That was more important to me than to go where the gold was. I wasn't too interested in that, as long as I had enough to eat."

After Chicago, New York was a revelation for Art. For the first time he found himself playing to audiences who came to hear the music. "In Chicago it was always the fan dancer, the cooch dancer, whatever, but *you* were never the feature. Same music you were playing and now you have an audience sitting there listening." Art looked up the Chicago crowd—Condon, Jimmy McPartland, Georg Brunis, Marty and Joe Marsala—and he began making new friends, including drummer Big Sid Catlett and Sidney Bechet. He even found himself accompanying vocalists again, Frankie Laine (whom he taught "That's My Desire") and Stella Brooks. Those whom Art played or recorded with during the dozen years of his New York residency constitute a veritable *Who's Who* of the jazz greats of that day. An abbreviated roster, in addition to those cited above, would include Barney Bigard, Rod Cless, Lee Collins, Vic Dickenson, Pops Foster, Chippie Hill, Bunk Johnson, Max Kaminsky, Mezz Mezzrow, Freddie Moore, Albert Nicholas, Tony Parenti, and Muggsy Spanier.

The critics discovered Art Hodes and wrote rave notices of him in the jazz press and the city's dailies; a young fan helped him secure a record date with Decca; he got gigs at Jimmy Ryan's and Nick's, was the regular host of a jazz show on WNYC, and for five years published, edited, and wrote for one of the first of the small press jazz periodicals, *The Jazz Record*.

Returning to the Chicago area in 1950, Art settled in nearby Park

Forest, a town of 35,000. From that base he has taken bands on the road from coast to coast and to Canada, visited jazz festivals here and abroad, performed at New York piano bars like Hanratty's, and played the odd gig in his old home town. "I made an armed truce with Chicago," Art joked. "I don't bother them, they don't bother me." He continued to record, has lectured on jazz at colleges, and co-produced (with Ed Thomas) a series for public television, "Art's Place." And he likes to "live a little bit of community life," playing weddings, benefits, and school assemblies, teaching his instrument, "shooting a little pool."

"I'm dedicated to my art and I'll do what I can for it," Art said, revealing a personal credo jazz artists have, for nearly a century now, collectively etched in stone, if you will. "And what influenced me that way was Jane Addams of Hull House, because all her life she stuck with her product, see, she didn't open a bowling alley when she got famous, she didn't open a riding stable. There's one thing when I was at Hull House that impressed me terribly, tremendously, and that was they had a sign over the theater—now, mind you, I'm a little kid walking down, see this sign that says, 'To thine own self be true. There all honor lies.' That made a lot of things easy for me."

Art Hodes, like Wild Bill Davison, was one of a handful of jazz artists still active through the 1980s who could look back to the teens and '20s and bring to life what the jazz scene was like then.

"The first time I met Jimmy Yancey," Art reminisced, recalling an encounter decades ago with the dean of Chicago boogie-woogie pianists, "was in the back room of a record shop. We were introduced to each other and he sat there and while he was sitting I started asking him questions. 'I understood that you were a dancer.' 'Oh yes,' and he told me about dancing before the king and queen of England of his time, and so I said to him, 'Jimmy, would you just do a step for me,' because I'm very rhythm-oriented anyway. So sitting there, he danced, he just moved his feet and—he sent me, I mean, I was really entranced, this man sittin' there dancin'."

We know from the testimony of Davison, Hodes, and many others that the white jazzmen spent much time in Chicago's Black Belt listening to the likes of King Oliver and Louis Armstrong. Some of them even sat in with their black idols, especially at informal after-hours sessions, for this was before the time of racially mixed bands playing for the general public, which did not occur for another decade. Hodes and Davison both spoke of pianist and bandleader Earl Hines with great affection.

Hines came to Chicago in 1924 and stayed even longer than Hodes did, for Earl's big band was at the Grand Terrace, the South Side's premier night spot, for a dozen years, off and on, beginning in 1928. This was Earl's initiation into big-time show business, for the band played for revues and other stage productions, much as Duke Ellington's band did at New York's Cotton Club. Like Hodes and Davison, Earl came into contact with the mobster element, and his attitude mirrors theirs in that he saw nothing and kept his mouth shut.

"I came from a musical family," Earl told me, recalling his early years in Pittsburgh. "My father played cornet and had a band called the Eureka Brass Band. My mother played organ, my uncle played all the brass instruments, and my auntie was in light opera. So I was surrounded by music. My mother and father both told me that at the age of nine they saw me watching her play, so they thought I would probably like to play the organ. Knowing that I wasn't able to reach the pedals as small as I was, they decided to get me a piano. Very few people had a piano and if you had one, you were really high on the hog. So they got me a piano. Incidentally, I still have that piano, it's still at my sister's—my sister has the same home we were in.

"My mother gave me my first lesson at nine and then we had a private teacher around there. I majored in music in school. I studied to be a classical pianist and placed all my heart and soul to be a concert master. But we had the racial trouble as to a black man being in that type of music. It looked like it would be an impossibility for me to get anywhere and I would probably be out of the picture by the time the racial situation was straightened out.

"So I happened to hear this tapping music going on at one particular time in a restaurant and I was *very* young and I said, 'Now that's something that I like,' and I went upstairs with my relatives. They called it 'syncopation'—it wasn't jazz, it was ragtime—and from then on I began to like it and that's the beginning of my career.

"My auntie was very attractive and all these big artists that used to come through Pittsburgh used to make over me so much in order to get to my auntie. And I was stealing and learning everything these guys were doing, taking advantage of the situation. After that I joined a big band. At that time we were playing upright piano—there were no grands—and we had no amplification. For singers they had megaphones. When I had a solo they could hardly hear me and I was trying to figure how I could cut through this big band. Well, I thought of my father playing trumpet—cornet at that time—so I started using octaves and playing a trumpet style, and that way I cut through that big band. I didn't do it for publicity or anything, it was just something that I wanted to do for myself. It got very popular and all the other pianists started doing the same thing. I was surprised to know that other people were doing it when I got to Chicago."

By the time he was twenty Earl had been a professional musician for five years. One of those artists he refers to as coming through his home town, and from whom he no doubt borrowed a few licks, was pianist Eubie Blake, almost forty at the time and destined to remain an active performer for sixty more years, almost until the time of his death in 1983, five days after his hundredth birthday.

"After hearing me play, Eubie says, 'You'll never get anywhere staying around a little town like this 'cause it's off the beaten path. I think you should get out of here. If you don't get out of Pittsburgh, I'll take this cane and break it over your head!' Fortunately, before he came back the following year I was invited to come to Chicago, and in coming to Chicago the avenues opened up for me. I started in a little nightclub and from then on I started to sort of stretch out and I did a lot of exploring. I was always a guy that wanted to have something on my own so I sacrificed going with different organizations until I got the experience to know how to conduct and handle an organization of my own."

Actually, Earl Hines apprenticed with several musical groups, including the orchestras of Carroll Dickerson and Erskine Tate, during his early Chicago period, working theaters and other venues with them and touring with the former for about a year. He became musical director and pianist for Louis Armstrong's group at the Sunset Cafe in 1927 and also worked with Jimmie Noone at the Apex Club that year. With Armstrong and Zutty Singleton as partners Earl was briefly involved in the club scene as entrepreneur at Warwick Hall. The project failed, but Earl's artistry was blossoming. In 1928 he made some 78 rpm sides with Louis Armstrong that, in critic-historian Gunther Schuller's estimation, for the first time established jazz as "having the potential capacity to compete with the highest order of previously known musical expression." Toward the end of that year he made a trip to New York to record eight solo sides for the Q.R.S. label.

"Before Louis passed," Earl recalled, "he said, 'We didn't know we were making history, did we?' And I said, 'We certainly didn't!' "

Earl's description of how those sessions went more than fifty years before our interview is a classic account of art unself-consciously coming into being.

"It so happened that passages that Louis was using were similar to passages that I was using, and then by, I suppose, having the same frame of mind, we became very close friends because we liked what each other was doing. We were using the record as a critic and we found out what mistakes we were making and then we tried to correct our mistakes by listening to the recordings. We were just recording, and ideas came for us while we were at the studio.

"I'd say, 'Louis, make this,' and he'd say, 'No, I don't want to make that because I don't think I got strength enough to make that.' This is what conversation we used to go through over there when we did those recordings. For instance, when him and I made 'Weatherbird,' he said, 'Let's you and I do a number,' and I said, 'Now what are we going to play?' And he said, 'I'll start playing sixteen bars and you think about another eight bars in the middle and I'll just follow you.' And that's how we made the tune up, just him and I. And we laughed about it when it was over, and come to find it turned out to be one of the hits. So you never know.

"Likewise, the same thing happened when I was doing the night broadcasts from the Grand Terrace. All over the United States they were running to the radio turning us on and I didn't know it until I finally started touring and people said, 'We heard you last night.' I didn't know I was making history as the first black band to be going network coast to coast. Little did we know the whole United States was listening when they said, 'You're on the air!' We just loved to do a broadcast."

Along with Fats Waller and Art Tatum, Earl Hines is a member of the pantheon of keyboard artists of the pre-modern period of jazz, and there are many who consider him the greatest of the three. Duke Ellington believed that the "seeds of bebop" resided in Earl's style of playing, according to jazz historian and critic Stanley Dance. Yet even if these distinctions did not belong to Earl, he would go down in history for still another reason. Arguably, Earl discovered more talent than any other bandleader in jazz history, certainly more diverse talent.

"I have to give the credit to the proprietor of the Grand Terrace," Earl insisted with irony. "Because he wouldn't pay any money for anybody, so what I had to do was *make* my stars. If it hadn't been for him, I probably wouldn't have discovered as many as I did. I began to see all this competition out there and I knew I had something to worry about. I couldn't go out there on tour the same as I did the year before. In traveling, I would always find someone who was advantageous to the band. That was my hobby, discovering talent, and I never will forget, the boss looked up at the Grand Terrace bandstand one night and he said, 'We got more musicians than we have customers.' At that time I had sixteen men on the stand.

"The first guy I picked up was Jimmy Mundy, a saxophonist and terrific arranger, and he finally ended up with Benny Goodman. Right here in Washington is where I got him from. He came over to the hall where we were playing and said, 'I have some arrangements I'll sell you for five dollars apiece.' Well, big band arrangements at that time were selling for seventy-five and a hundred dollars, so I was a little doubtful of this guy. I rehearsed one of his numbers and I said, 'Good gracious, this guy's way ahead of his time.' I grabbed him and took him back to Chicago."

A very abbreviated list of others whose careers either began in or were enhanced by membership in bands or combos lead by Earl Hines would include singers Billy Eckstine, Johnny Hartman, Sarah Vaughan, Helen Merrill, Etta Jones, Herb Jeffries, and Marva Josie, trumpeter and violinist Ray Nance, trombonist Trummy Young, clarinetist Omer Simeon, and saxophonist and arranger Budd Johnson. The Hines band of the early 1940s was really the first big band of bebop. For most of

1943 Charlie Parker was playing tenor saxophone in it and for a few months in the early part of that year Dizzy Gillespie was also a member of the orchestra. But we are stepping into another period with those last observations, a period that will be dealt with in a subsequent chapter.

A few months before his death at seventy-nine in 1983 Earl was on the road with a new combo that he had put together in Oakland, California, which had been his home since the 1950s. "I didn't know that there were that many good musicians there," he told me on a visit to Washington, D.C. "These kids are very talented and I think they might make a good reputation for themselves." Earl had been ailing off and on for half a year or so and he confessed he was "not supposed to be out now, but I told the doctor, gee whiz, I gotta do something." He continued to play until the weekend before he left this world.

"This profession came down biblically," Earl mused. "During the days when the country was in trouble the king used to send for the jester and the musicians to make the people forget about their trials and tribulations. And that's been handed down. For instance, when you make an appearance before an audience, you're making an appearance before some people who had a disappointment at home, some people had a bad day at the office, and some people had a divorce, some trouble with themselves as far as illness is concerned, and all these different kinds of people come out to the theater or the club to forget about their everyday disarrangements. They want to come to some place where they think that they can enjoy and put all this behind them. When I'm out there I'm trying to reach the audience and find out exactly what they want and the first thing I do when I walk out there is let them know that I know what I'm doing, and by the time I'm through they've forgotten everything they've ever had happen, and I've yet to have somebody say they didn't like the show. People say to me, 'You guys are having a good time up there, aren't you?' And I say, 'We're having a ball!' "

Notwithstanding Art Hodes' dismissal of the Chicago musical scene as a "ghost town" in the 1930s, much was happening and continues to do so to this day. True, many of those whom Art had come up playing with had gone on to New York and elsewhere, while he remained in Chicago "still chasin' the blues." He had good reason for doing so, for the blues remained, and still remain, prominent on the Chicago musicscape.

As for the continuing evolution of jazz, we'll hear about the important role Chicago played during the 1960s as a shaper of things to come when we come to the chapter on post-bebop developments.

Pianist Dorothy Donegan, a native Chicagoan some twenty years Art's junior and a child prodigy, recalled the 1930s in vivid language. "I used to play house rent parties on the South Side when I was twelve or thirteen," said Dorothy. "That was for twenty-five cents an hour. I could stomp and keep time and hit the piano real hard because I played baseball in the alley. Then I used to play in a place called the All Stars on 43rd Street and they had cuttings and shootings in there. I used to play for two dollars—that was a lot of money. Art Tatum came to my house when I was seventeen and after he heard me play he said that I was the only woman that made him practice." That last claim is not braggadocio, for Dorothy is considered by some authorities to have pianistic skills second only to the great Tatum.

Another native Chicagoan, pianist Junior Mance, is four years younger than Dorothy Donegan and has similar memories of the 1930s. "Oh, I remember rent parties," he told me when he performed at Blues Alley in Washington, D.C., in a Memorial Day tribute for Count Basie, Red Garland, Juan Tizol, Mabel Mercer, and Z. Z. Hill, all of whom had recently died. "More or less a party where they would charge admission to raise the rent money for someone. They used to have a boogie-woogie pianist or a stride piano player, someone to just play the blues all night, and they illegally sold whiskey and food. At the time I was coming up in Chicago, that was basically *the* blues town. Most of the blues singers from the south stopped in Chicago, instead of going to New York. My mother and father were very fond of the blues and boogie-woogie and my father played for his own enjoyment, sort of a cross between boogie-woogie and old-time stride piano. They used to buy all these 78s and I used to listen to them. The first music I played on the piano was boogie-woogie when I was about five years old, almost like Sugar Chile Robinson."

Still another pianist, Andrew Hill, a Chicago native, gives us a sense of the transition from the earlier forms to the modern idiom. We talked several times at length about his background, and on one of these occasions he expatiated upon his coming-up years.

"In Chicago at the time," Andrew explained, zeroing in on the late 1940s and early '50s, "music was in all the neighborhoods and you grew up in a tradition of hearing music all the time." A precocious youngster, Andrew began piano early and along the way picked up tap dancing and singing and learned several other instruments, including organ, accordion, and baritone saxophone. He combined all of these skills in talent show appearances and once won two Thanksgiving turkeys at a Regal Theatre contest sponsored by the *Chicago Defender*, the black newspaper that he used to sell on the streets.

"I was able to start my professional music career at age fourteen with a job with Charlie Parker, and from my late teens to my midtwenties I played with every professional jazz musician that was alive," asserted Andrew with, we must observe, no little hyperbole. However, the roster of his associations is an impressive one for, to name only a few, it boasts Dinah Washington, Ben Webster, Oscar Pettiford, Miles

Davis, Kenny Dorham, Eric Dolphy, Ira Sullivan, and Bobby Hutcherson, constituting a spectrum of a sort that few jazz artists are capable of spreading themselves across. Yet for Andrew Hill this poses no problem. Andrew is a musical visionary who is steeped in the blues.

"I had a series of talks with Bird that night," Andrew said of the evening he spent with Charlie Parker when he (Andrew) was hardly into his teens. "He was telling me there will be a time in my life when I will be a keeper of the flame, simply from having experienced certain areas of music in the black tradition, things that only I and a few others really know about."

Indeed, as we shall see in later chapters, a number of such "keepers of the flame" were there to stoke the fires of the new directions jazz would take in the 1960s, '70s, and '80s in New York, Chicago, Los Angeles, and other locales.

3

Kansas City and the Southwest

We hadn't eaten in a couple of days and nothing was said, because the music was our survival.

Mary Lou Williams

I think the fact that you have to dig for it yourself, if you've got any talent, that'll make a stylist out of you.

Herb Ellis

I think as long as people don't turn into computers, and have emotions, and look to their emotions, there will be blues.

Barney Kessel

Along about this stage of our informal account of the evolution of jazz I sense a need to clarify something. A useful analogy, one that is quite familiar to me in its details, comes to mind, namely the year-long survey course that any institution of higher learning worthy of the name must include in its catalogue, History of the Ancient World.

The Story of Western Civilization, from its beginnings to the collapse of the Roman Empire, I found upon preparing an outline the summer before I taught the course, could be most handily compartmentalized. There were the Sumerians, Egyptians, Phoenicians, Hittites, Etruscans, Persians, Lydians, Greeks, Romans, and some others, and one could make some references to non-Western developments along the way. All very neat. Yet I found myself constantly backtracking, and if an inconsistency had not come to my attention, be assured that it would come to one or another of those sitting in (I liked to think) rapt attention to my lecture. There were always, it seemed, those little threads to tie up, this influence to deal with, that overlapping process to account for. For each of those cultures continued apace through some, in most cases all, of the subsequent period dominated by the next power to come to the fore. So it wasn't so neat after all.

Somewhat the same circumstances prevail in the case of jazz. The music did not grind to a halt in 1917 when the Storyville bordellos were declared off-limits to U.S. Navy personnel and closed up shop, thus convincing a number of jazz players that they would be better off seeking work in, say, Chicago, which, rumor had it, was wide open, or California, where they had heard that the sun shone and that few there had yet heard their brand of music. Yes, all that happened, but the music continued to happen in New Orleans as well, and still does to this day, from the oldest styles to the newest.

Nor—*pace* our dear friend Art Hodes—did Chicago become a "musical ghost town" in the 1930s because Eddie Condon and the rest of the "Austin High" crowd, Bix Beiderbecke, Louis Armstrong, even several of the more prominent boogie-woogie pianists, and many others departed for New York, went on the road with big bands, moved west, returned to New Orleans, or opted for a warmer clime than the Windy City. While Chicago would never again see anything like the frenetic action of the '20s, the music retained a foothold there and the city remains an important jazz and blues center. It has for several years hosted a mammoth free municipal jazz festival that features impressive local talent along with imported artists from here and abroad.

All of that by way of introduction to yet another one of those neat, handy categories that serve, in truth, only as that—a manageable way to deal with a subject of immense complexity, a complexity that jazz shares with any other art form that comes to mind. However, keep in mind that the evolution of this music is compressed into still less than a century. Those ninety-five or so years since—eye- and ear-witnesses have testified—the cornet player Buddy Bolden played the streets, cabarets, and dance halls of New Orleans, jazz has undergone changes that are no less than astonishing.

Some of those changes—the incorporation of country blues into a piano style that came to be known as boogie-woogie, the increasing emphasis upon the soloist, a contribution made in the main by Louis Armstrong, and the creation of Chicago Style by young whites of the city—came about during the teens and '20s in Chicago. Now we must turn to another part of the country: the Southwest and its musical center Kansas City. We shall learn of no less startling changes for which this area was the incubator. Once again, let us allow some of the musicians who came up during this period to tell us what it was like. After all, it is their story.

"I was born in 1908 in Honey Grove, Texas," pianist Sammy Price told me, "and by 1912 I was really getting around listening. A man came to Dallas from Athens, Texas, and got me to play for a house rent party for two nights, Friday and Saturday, the 23rd and 24th of December 1923 and I got five dollars for that. I got back in time for Christmas dinner, Sunday. That was my first professional job. Boogie-woogie came out of wherever there were black people. Blind Lemon Jefferson used to play those movements on his guitar and they called it 'booga-rooga.' They also called it 'fast western' but that name didn't catch on." The idiom was also played in mining camps and lumber camps, Sammy pointed out, "and you had to be a rough and tough kind of guy to go around those '49' camps."

The observations of Sammy Price were supplemented by an account provided me by the blues pianist and singer Sunnyland Slim (Albert Luandrew) in an interview between sets at Washington, D.C.'s Childe Harold, a restaurant that sometimes featured blues artists, including such legends as Sleepy John Estes, Professor Longhair, and Sunnyland Slim.

"The real foundation of my life," Sunnyland commenced, "I was born out in the country [in] a little old place called Vance, Mississippi, but at the time I was born it was country, town, farm, corn, cotton, all was farmin'. My peoples was preachers, my father was a preacher, my grandfather was a preacher. After I got up seven years old or a little before that, my mother died and I stayed with my grandfather. At this particular little place—Vance—we lived out in the country. My grandfather split rail for this old place and bought it back in the last of the 1800s. And I was raised on this farm up until the age of around eight. My father married again, that's what caused my life to drift.

"Long about the time my father married he bought another place. And when I got old enough to listen at the gui-tars—we didn't have organs in every place—I started going to this little town called Vance a little further down the road and I heard this piano." Sunnyland here alluded to a "barrelhouse" as the place where he was first exposed to the piano style that would become one of the elements he would incorporate into his own approach to the blues. "That always stayed in my ears," he pointed out. "Then they carried me to a minstrel show and that stayed in my ears, the horns and things like that.

"My stepmother was so mean to me," Sunnyland recalled, "but I was too little to run off, too young, you know, so I went up to my grandfather, the other farm. . . . When I got up about thirteen I run off, went to Crenshaw, Mississippi, and they found me and brought me back. One of the mothers of the church where my father was pastor, she had an organ. If it rain, I could get a chance to cut the woman some wood, bein' her husband gamblin'. She had me cleanin' house and I'd cut her kindlin' wood and her husband'd be gone and I asked her, 'I want to play your organ, I cut you some wood.' I practiced on it, tried to make what I could.

"There was a boy named Jeff Morris, he played piano. I would learn a little bit on the organ and show him. I'd do it my way and what he showed me I made on a shoe-box top. And I'd get that sound and carried it out there and made a F chord. By me bein' a choir member, singin' in church, you know, I would learn to make that sound. I could hear that little march, that trumpet sound. But I didn't know nothin' but F. Another boy out at Becker played one of those little old funny pieces, but I didn't like this style of music too much. I liked the uptown music, I liked the way they played, them horn players.

"So I hung around with this lady and after I run off another time they couldn't get me. And here's what sho' nuff got me into findin' out about playin', learnin' how to play good. I learned how to play a little in C and F, I learned a little C, but what really got me over—I went to Midnight, Mississippi . . . and to Canaan, Mississippi, and I run up on Little Brother Montgomery."

Sunnyland Slim's initial encounter with Little Brother Montgomery was at a logging camp. "The camp was so muddy," Sunnyland began, "a great big camp, great big upright sawmill, upstairs sawmill, and that's where I met Little Brother Montgomery. He was playing very beautiful, . . . the blues and little boogie-style things, 'Shim-me-sha-wabble,' and 'Black Bottom.'

"It was 1923 when I met this cat and I was about broke, I had a dollar or somethin' and I come in there where he was and he had a glass on the piano. He didn't know nothin' about me and I say, 'Hey, pardner, could I sing one?' And he said, 'Can you sing?' I said, 'Yeah, I can sing. Play me some blues, 'cause I got 'em. My car broke down on the road.'

"And I went to singin' the blues and he played and the peoples got up out of the bed. It was mornin', before day, about four or five o'clock, and I went to singin' the blues and the womens come up with hip boots. Of course they got to come through the mud and come in downstairs. It was so muddy around there, one of them sawmill camps, wasn't no sand, no gravel, like it is now, you know, 1923.

"And everybody got up in that camp," Sunnyland recounted with great pleasure. "I'se walkin' the floor singin' the blues. And I made about two dollars, the peoples gave me, and the man gave me a nice meal, and I stayed around there until about the next night. That was a beautiful time ever was, you know, and that's when I met Little Brother Montgomery. Later on he got drunk and I played a couple pieces, and I felt kinda at home."

"I was born in Sherman, Texas," multi-reed player Buddy Tate began an interview during intermission. Buddy, who was born in 1913, was in a week-long residence with tenor saxophonist Scott Hamilton at the Maryland Inn, Annapolis. "At the beginning we had a family band, not brothers and sisters but cousins and what have you, and we worked it up to be a pretty popular little band. We were playing for all the colleges. In my town at that time, as small as it was, we had about five or six colleges. I was about thirteen years old. My first instrument was alto. We had a boy from Dallas and his name was Milton Thomas, a very, very good saxophone player and he was playing alto at the time and I loved him. My brother played C-melody saxophone. The pianist, he didn't play professionally then, but the one that played C-melody did. It was so much difference in our ages, like he was eight years older than me and the pianist was ten years older. That makes quite a difference when you're in your teens. I played with the band all while I was going to high school, and we were making nice money.

"There were a lot of blues singers then, like Lonnie Johnson, and that was one of my first gigs. I gotta tell you about that. That was the biggest money that I'd ever made. Lonnie Johnson was a big man, you know, he had hits one after the other. Between my home town and Dallas, which is now about forty-five miles but it was farther then— they've cut roads through—there is a place called by the name of McKinney and my aunt ran a dance hall in McKinney. Lonnie Johnson was playing a dance there and a concert, like, and when he got there that afternoon about, I guess, five or six o'clock from Dallas his band didn't show.

"Now there's nobody in McKinney can play anything so he gets worried about seven o'clock and he goes and asks my auntie, he says, 'You know, I'm worried about my band, I don't know if they're going to get here.' He says, 'Have you anybody here that can play?' She says, 'Oh, no, no musicians here. Oh, wait,' she says, 'I've got some nephews up the road.' That's the way they speak down there. 'Up the road' would be a hundred miles. He says, 'Well, can they play?' She says, 'Well, they *say* they can. They play for dances.' He says, 'Well, if they can play for dances that'll be find. Can you call them?' She called us. It was about twenty-five miles. I would say that we got there in about twenty-five minutes. To play with Lonnie Johnson! You *know* we were all excited.

"That was fast traveling and we made it and when we got there he said, 'You guys play dance music?' We said, 'Oh, yeah.' Now that was our thing so we weren't nervous when he said that. Then he says, 'I'll only do about three or four numbers and then you play for the crowd.' They were all lined up all around the corner. So we went on and did some blues and then he turned it over to us to play some dance music and he loved it, fell in love with the band and everything, because we had it well rehearsed.

"So when we finished playing he called me over in the corner, he says, 'Come on now, here's the money,' and he started peeling it off. He gave me fifty dollars, which was nice, you know, that was a lot of money in those days. There's five of us, ten dollars a man, a real nice big deal. So I called the guys over and says, 'Come on here and get paid.' Lonnie Johnson says, 'What are you doing?' I says, 'Well, I got to pay the cats.' He says, 'No, man, that's yours!' I couldn't believe it, I thought the world was coming to an end. 'It's yours,' he says. 'I'm going to give 'em all fifty dollars.' And I said, *This* is *it,* there's no business in the world like this.' And this is really true, all right, that he gave us fifty dollars apiece.

"At that time I must have been about fourteen or fifteen. I was still going to school and we were playing all the neighborhood towns and things and were drawing good crowds. I imagine it was about three, between three and four years later, that I went professional. I left, me and my cousin, and went to Wichita Falls, which was an oil town and there was just plenty of money out there, and Vernon, Texas, Jack Teagarden's town. He was playing in the oil fields. We went out there and joined a band and that's where Herschel Evans and I worked together, in Troy Floyd's band. That's when I started playing tenor and Herschel was playing lead alto. I was in that band for a while and then I joined T. Holder's band. I worked in Holder's band off and on for three or four years, playing with other, professional musicians, a lot of traveling and meeting musicians, and things. We would go from Texas to as far as Kansas City and even to Chicago, which was kind of a broad kind of a big territory, but Holder was sort of famous, you know.

"In those days you had bookers and they would book certain territories. Maybe you would play in a radius of, say, two hundred miles. The way we did then, we would take all the expenses out, take all of that off the top. Everybody shared in paying the bills. So if you had a twelve-piece band, the booker would be the thirteenth man, break it down thirteen ways. Made a lot of money that way.

"I've traveled five, six hundred miles, traveled that far and played. Like we finished playing and sometimes we'd leave right after we finished playing a dance, maybe drive four hundred miles and play that next night. That was kind of rough. Change your shirt and go on out—oh, yes, I've paid those dues. Sometimes you wouldn't have time *to* change."

The key word several paragraphs back was "territory." A territory band, precisely described by Buddy, is one that more or less stays within a prescribed territory, in those days usually no more than a four- or five-hundred-mile radius, playing one-night engagements for the most part. The West and the Southwest were full of such units in the '20s and '30s. Various cities—Omaha, Little Rock, Oklahoma City, Dallas, Kansas City—served as home bases for one or another, perhaps several, territory bands that were often on the road for months on end meeting the demand for dance music. As time went on some of these bands—Andy Kirk's and Count Basie's being prominent examples—gained national fame and toured much of the country. And that story—the story of the big bands—will be told in a later chapter. For the moment, let us check in with several of the principals of the Kansas City scene.

"I first heard the blues when I was a young kid in Kansas City, Missouri," is how singer Big Joe Turner commenced an interview between sets at the Showboat in Silver Spring, Maryland. "I used to listen to Ethel Waters a lot, and Jimmy Rushing, Hot Lips Page, and, let's see, Johnny Walker, Blind Lemon Jefferson, and a whole bunch of cats. I listened to people play on the streets. I used to lead a blind man around—who was that? I can name you anybody but I will be maybe wrong. His name was Johnny Creech and he played a pretty fine guitar.

"I used to be a bartender and I learned how to sing by going up to sing with one of the bands in the intermissions. They had two bands at the Cherry Blossom. I used to have a lot of fun with Pete Johnson. I met him during the time when I was a kid. He was a piano player that played in the Hole in the Wall. We had a lot of cats drop in. Count Basie's boys always fell by and different people who worked around the different nightclubs would come in and have a good time. So we'd have a jam session—Joe Williams, Jo Jones, Hot Lips Page. There were so many people that dropped in there, I don't remember all the people. We had Louis Armstrong, Count Basie, and all the different fellows come through there and we had a ball. We never started the jam session until about two-thirty and we'd play on to five or six in the morning. Everybody'd be coming from different clubs where they worked at, they come by and played. We'd fill the house up every night. My voice was so powerful I'd sing without a microphone. We didn't use no microphones.

"I was doing rock and roll music way before they even called it that. I always been doing it, it wasn't a brand new thing, rock and roll was just a new name for it. I was doing that stuff all along, doing my thing." Despite failing health in the final years of his life Joe continued doing his thing until very near the end in 1985 at the age of seventy-four.

Sammy Price's observation reinforces Joe's claims: "If you listen to Jerry Lee Lewis, he'd tell you he created rock and roll. I don't mean to discredit those guys—I think they're talented—but I don't think that they should take the credit for something that was created as an art form in the black community."

For Jay McShann, arrival in Kansas City was on the order of a baptism by fire, especially considering how he had grown up in Oklahoma. "My folks was quite religious and they gave my sister lessons,"

Jay revealed to me on a visit to Charlie's Georgetown, a Washington club, "but they didn't want me playing any of that 'devil's music' on the piano. So, quite naturally, they grounded me from playing the piano. The way I would do, I would tell them I was sick on Sunday. Everybody goes to church. I'd just sit there and enjoy myself, until one day one of the church sisters came by and said, 'Did you all go to church today?' My mother said, 'Well, everybody went to church except this boy—he had a stomachache.' That church sister said, 'Well, honey, I passed by and somebody was sure playin' them reels!' So, boom! I caught it again."

Here is what greeted Jay when he arrived in Kansas City in 1937 at the age of twenty-eight: "I hadn't been around nothing like that. They had music piped out in front of the clubs. I was just going in and out of every club. I didn't want to miss nothing, stayed up day and night. There was Joe Turner and Pete Johnson—they'd probably do one tune and it would last for an hour. I'd never heard no boogie-woogie and blues like that, see. Pete would roll 'em for about twenty minutes and then Joe'd sing about thirty minutes. The clubs stayed open around the clock. Some musicians go to work at eight o'clock, eight to five in the morning. Then the jam sessions took over. There'd always be one club where they'd have a 'spook' breakfast and everybody'd go to this club after all the musicians were off, like five or six in the morning. You might say, 'Where's the spook breakfast going to be this morning?' And they'd say, 'It's over at such-and-such place,' and everybody'd go there and play, and that goes on until twelve o'clock in the day. And when the cats come into town, there's always a session going on and that made it just that much more interesting."

Mary Lou Williams's career, which included periods of residency in Kansas City, was a full one. As pianist and arranger for Andy Kirk's Clouds of Joy, formerly T. Holder's band, during the 1930s Mary Lou traveled widely and made many contacts, which she followed up on as the years progressed, eventually contributing arrangements to the books of Benny Goodman, Earl Hines, Duke Ellington, Louis Armstrong, Tommy Dorsey, and others. Her home in New York became, during the 1940s, a salon where the young beboppers would hang out on a regular basis, trying out their latest licks. Her final gig was as artist in residence and professor of music at Duke University in the late 1970s, and she died there in 1981 at the age of seventy-one.

"You see, I was well trained," Mary Lou pointed out to me in an interview held in the studio of WGTB-FM at Georgetown University. "I began playing at the age of three and at the age of six I was playing for the entire block where I lived in Pittsburgh, playing for the neighbors, receiving money. I was trained by older musicians and I listened to them. They always trained me to play all styles and play everything, and so in my early years I listened to the records of Jelly Roll Morton,

Earl Hines, James P. Johnson, and Fats Waller. When I was about the age of twelve I was playing engagements with the bands. Some of the bands would come to Pittsburgh and stop by the house and take me out on their gigs.

"In the beginning I couldn't read music," Mary Lou explained. "My mother used to hold me on her lap, she practiced the organ with me sitting on her lap, because seems like I'd get into a lot of trouble, breaking dishes and whatnot. One day she was pumping this old-fashioned organ and I began playing it. It must have been terrific because she dropped me on the floor and ran and got the neighbors and brought them in. Then in the second grade I discovered that I had perfect pitch. My teacher lost her harp and so I said, 'I know the note you started on.' She went to the piano and I was right."

Mary Lou Williams participated in the all-night jam sessions in Kansas City and on one occasion was roused from sleep in the middle of the night to serve as relief for an exhausted pianist. Her accounts of life on the road with the Clouds of Joy, of which she was a member from the early 1930s until 1942, are quite harrowing.

"During the years I was with Andy Kirk we starved almost. I remember not eating for practically a month several times. But we were very, very happy because the music was so interesting, and you forgot to eat, anyway. Everything was laughter and we had a great time. During the Depression we played engagements and we knew we weren't going to get any money because Andy would scratch his face when he was walking toward the band and the trumpet player would pull out his horn and play the 'Weary Blues.' And we'd laugh about it. We hadn't eaten in a couple of days and nothing was said, because the music was our survival, I think."

How far the music can go in terms of helping people survive was given example in a recollection of Mary Lou Williams, who was many years ago dubbed the First Lady of Jazz. "I had a guy come to the club where I was working in New York who was on the verge of suicide. He wrote me a fan letter. It said, 'I walked by the Hickory House and saw your picture and just went in and sat down and listened to you. Now I can really make it and I was on the verge of committing suicide.' You see, the music has healing powers in it. Everything you talk about is in it, healing the soul and all that, and people have forgotten to listen to it as a conversation. If you listen to it you'll find that it has a story."

Up to this point, we have concentrated in this chapter on some of the black musicians who came up in the 1920s and '30s in the Southwest, and we have given a prominent role to Kansas City as the principal hub of activity for the territory bands of that period. For several reasons Kansas City was exactly that.

For one thing, Kansas City was the commercial center for the Southwest, and its reputation as the region's cultural center, as well, was unchallenged during the peak years of jazz action there, the 1930s. For another, the city was wide-open in terms of opportunity for night-life because it was controlled by one of the most corrupt political machines in the history of our nation, the Pendergast organization, which allowed gangster elements to take over the entire industry of alcohol beverages, from manufacture to retail sale. The city was, therefore, an ideal environment for another stage in the evolution of jazz.

As for the white players of the Southwest, they too participated in this evolution, for the white bands that were booked at various dance halls of Kansas City and other urban locations, large and small, throughout the region included the Coon-Sanders Nighthawks, the Casa Loma Orchestra, and the bands of Hal Kemp and others. Some of these were certainly not full-fledged jazz bands but had jazz soloists and played some jazz arrangements.

Two white guitarists who were in their teens during the second half of the 1930s have interesting stories to tell about the influence of the Southwest on their styles and upon their later careers, which have for some years been acted out in the international arena. Herb Ellis was born in 1921 in McKinney, Texas, the town where Buddy Tate's aunt ran the dance hall where Lonnie Johnson came sans band about 1928 and hired the fifteen-year-old Buddy and his friends as back-up. Barney Kessel grew up in Jay McShann's hometown of Muskogee, Oklahoma, where he was born in 1923.

"I've really never tried to analyze it too much," was Herb's initial response to my query, How did he learn to pick the blues with such feeling? "I really don't know, but I'll try to probe my mind and see if I can come up with something, sort of try to evolve something, make something up—not a lie, I mean, see if I can figure that out.

"I think in my case it may be the section of the country that I was raised in has something to do with it. I think really the actual setting of where I was raised, my surroundings, the home life that I had, all had a part in it, I believe, because I was raised near Dallas, but way out in the country, I mean *really* in the country. Why I say that is, you know, there was no neighbors around for a half-mile or so. This sounds like a really sad story—I don't mean it to be like that, I'm just trying to equate. I had nobody to play with, had great parents and a brother and a sister. It was a very wholesome but lonely existence. You can drive out there to this day and look at that part of the country and you feel lonesome because it's just lonesome and desolate. I had bluesy, lonesome feelings all my life when I was a kid. At night I'd just sit out on the front porch and hear the harmonica playing in the distance and it would bring tears to my eyes. That feeling, someway, gets into my music. It *must*, because I feel certain emotions when I play blues."

Barney Kessel also had early exposure to the blues idiom. "Being from Oklahoma, I started playing, really, at the age of about thirteen and a half, with a thirteen-piece all-black band and they played a lot of blues. However, their blues mostly were very happy blues, and I didn't really know that there were sad blues for a long time after that because everything we played had a beat and it was very upgrade. But it was called blues. The other kind of blues that I began to be aware of was that I used to go down to the railroad tracks and there would be boxcars, I mean trains would go by with boxcars, and there'd be any number of hoboes. I saw many one-man bands, guys with a kazoo and a guitar and a washboard, at a time when they weren't in show business and people were wearing blue jeans because that's all they could afford. It wasn't the hip thing to do, they were just living that way, and I saw all these people and a lot of them were playing blues. It was living in Oklahoma where there was a lot of that kind of music at the time that I grew up that led to this feeling that stayed in my playing."

The conversations I had with these two master guitarists were in the same hotel room at the same time and it was most interesting to bounce the same questions off the two of them. I couldn't resist inquiring whether they had any hope that the blues would stay around for a while as a genre. That is, recognizing that the sorts of environment that had shaped the blues feeling in both had largely disappeared, would it be possible for later generations to learn to play the blues?

"I think one can learn to play the blues in the same way that one can learn to be a French chef, by looking at a book, or one can learn to become a bullfighter by taking a correspondence course, or one can adopt the culture of a gypsy reading books on it," Barney began. "I don't think it's an impossibility, but I think to do it that way would result only in an intellectual attainment. I don't think it would be an emotional one at all. I think your experiences in life lead you to develop within yourself a feeling that is highly empathetic toward the blues. It isn't a matter of turning a blues button on like you're a computer and saying, instead of playing rock or folk or this and that, 'I'm going to play the blues.' I think it's a feeling that comes to you that mixes with the music and if you haven't had that kind of experience, it's not going to be there. I don't think you're going to learn it from Berklee College of Music. All you'll learn there is how other people do it, but it's without the emotional content, which we never talk about in any scholastic training. You don't go to any school and enroll in a course with a teacher and get a grade on emotion. This has to come from you and your life experiences. I think the blues can be taught, but only the intellectual part.

"As far as Will the blues continue?, I think it'll continue as long as human beings retain their feelings and have feelings of despair and desolation and feelings of blues sometimes and even the feeling of the joyous part of the blues, like 'Goin' to Kansas City' or something like that, where it's on a positive upbeat, that kind of blues. I think as long as people don't turn into computers, and have emotions, and look to their emotions, there will be blues."

"I feel, as Barney said about the intellectual part of it, you can teach the mathematics, but you can't teach how to play it where it'll mean anything," was Herb's immediate response. "It's got to have a personal value in it, personal experience, and I think there will be other blues, certainly, the music is not going to stop, and there will be blues players coming up. I don't think the blues are going to die out, not in the foreseeable future. I would wonder what kind of blues players we're going to have. From my point of view, I don't think you're going to get any really meaningful blues players from the angry part of society 'cause you're going to get some angry blues, it stands to reason. There may be a lot of meaning in it, but it doesn't really say too much to me, 'cause I don't think music is a point to vent anger by. That's just a personal feeling, it's not my cup of tea. But you're still going to get some good sound, some meaty blues players, 'cause they do reflect society, and you're going to get some disjointed weirdos playing some blues that you could put 'em all in a can and ship 'em out to the ocean, that's the way I feel," Herb concluded, laughing.

Finally, let's hear how these two gentlemen began on their instruments as youngsters in the Southwest in the mid-1920s and during the Great Depression.

"I started not on the guitar, I played the harmonica at first," Herb revealed, "which I hardly remember because I was three years old when I started to play. Some people remember that part of their life but I don't. I remember having played it a little later and then 'my sister got a banjo for me to instill more musical aptitude in me, not aptitude but hopefully make something out of music, and I liked it all right because that's the only instrument outside of the harmonica I had. Then there was a guitar left at my house, just left there, and I just picked it up. I learned how to tune it and I ordered a book [that] told you how to play the guitar. I was all self-taught, I never had any formal lessons, it was all just by getting all the knowledge I could and all the information about guitar playing that I could get at that time, which was hard to come by, really. You didn't have too much to go by then. All you could just do is listen to the radio and get a few sad books with a few chords in them. You had to get it for yourself, which I guess is good in a way, now that I think about it. When I did go to college I majored in music, but I did not study the guitar because there was no guitar teacher. So I've never had a lesson.

"I'm not saying, isn't that terrific? I'm just saying that's the way it is. Maybe in my case it *is* better, I don't know, I wouldn't have any way of knowing. I do know one thing, I know Tal Farlow didn't either, not really. I mean, he had a few lessons from the WPA or whatever it was, but he's self-taught, Wes Montgomery's self-taught, Joe Pass is. Why I mention that, all individual styles, all definitive styles, and I think the fact that you have to dig for it yourself, if you've got any talent, that'll make a stylist out of you, it'll separate the chaff from the wheat. If you've ever had your wheat chaffed, you know how painful that can be," chuckled Herb, turning the discussion over to Barney.

"That's right. I started playing guitar at the age of twelve, and at the very time that I started I had lessons at that point and they were of three months duration. They were part of the WPA (Works Progress Administration), the program called the Federal Music Project. It was during the summer months when I was out of school and it was four hours a day, nine 'til eleven and one 'til three every day, six days a week for three months. As I look back on it, there's nothing that I would change. I could not today, if I were giving a course on guitar, give any better introduction. There was nothing to unlearn, nothing to relearn. I started out by learning guitar chords, what each note was on the finger board, and as I played the chord not only what was that note but what part of the chord it was. I started learning that at twelve in a class of about thirty-five people.

"After that I never had any more lessons in my life, although after playing about fifteen years I got very, very curious to learn some things and started to study privately with teachers, studying arranging, composing, conducting, went to UCLA and took a night course on writing music for film, but that was like to educate myself and *some* of that I was able to apply to the guitar.

"Otherwise, it's just like Herb, it was a matter of a *lot* of experience, many, many, years of experience, and a constant hunger to educate yourself, listening, learning, just more or less always asking, and something that we don't really talk about too much, that should be said, you have to have talent. Talent, training, experience, perserverance, good luck—you have to have all those ingredients."

In this chapter we have touched upon some of the elements that came together in the fusion loosely known as Kansas City Style. Blues and boogie-woogie, the territory bands, and the jam session were clearly principal constituents of the general style of jazz that came into being in the Southwest. We could add that St. Louis and Kansas City were important ragtime centers from the turn of the century until the idiom peaked in the late teens. There is also some evidence that New Orleans musicians, including Jelly Roll Morton, occasionally passed through the area. So, as remote as Kansas City and the Southwest were from the beaten path, the main components of jazz as it existed in the 1930s were nevertheless present there. Later developments such as the big band sound, 52nd Street cutting sessions, Jazz at the Philharmonic type free-for-alls, and rhythm and blues can all be said to owe a great deal to the Southwest in terms of their origins. One direction of the music we have not yet made mention of here is bebop, routinely thought of as having leaped into being in New York in the early 1940s. But if you lend your attention to the following account of a 1943 incident, you'll see that it too had deep roots in Kansas City, where Charlie Parker grew up.

"I was with Tiny Brandshaw and we stopped in Kansas City," Sonny Stitt told me between sets at Blues Alley. "Next to the hotel was the union, so I threw my luggage in there and then I took my saxophone and went down to 18th and Vine. I picked him out of a crowd of people. I don't know how I did it. Well, I'm clairvoyant, anyhow, I think a little bit, you know. Anyway, I said, 'That's him.' I say, 'You Charlie Parker?' He say, 'Yeah, I'm Charlie Parker, who are you?' I say, 'I'm Sonny Stitt.'

"So we went on to a place called the Gypsy Tea Room and we had a little, brief, impromptu jazz session between ourselves with a piano player. He said, 'You sound like me.' I said, 'You sound like me, too.' So we got to be good friends and he was always a gracious man in my life. I was his pall bearer, man, and it broke my heart. I wouldn't even work that week for nobody.

"I'm going to tell you right now, I don't know anyone that can match him or beat him. I come a close second," Sonny concluded with a laugh.

Having made the point that Charlie Parker, too, came out of the Southwest and brought some of the seeds of bebop with him when he first came to New York in 1939 and began participating in the jam sessions at Monroe's Uptown House and other incubators of modern jazz, we move on to New York in the next chapter. But first let us emphasize that a great many jazz artists besides those mentioned in this chapter came from the Southwest, players who were swept up into the swing and bebop big bands and combos, as well as, a few years down the line, into later-developing styles of the music such as the free form that came into being in the 1960s. Note, for example, that none other than Ornette Coleman came to New York via California from his origins in Fort Worth, Texas. And keep in mind the remarks of Professor William Schafer (in our opening chapter) expressing skepticism that one can reduce the history of this music's evolution to "a very neat sort of map that you can draw on a blackboard." Such a map simply does not exist for tracing the roots of Afro-American musical idioms.

4

New York

You could walk into any of a dozen clubs in Harlem and hear some great jazz in 1925, '26, and '27, along in there.
                                    Spencer Clark

All of the artists that I had been reading about in magazines and whatnot . . . I got a chance to see because just about every known musician in the business was in New York.
                                    Milt Jackson

New York is it, always has been, it's the mecca, the testing ground, and I can't see any other change.
                                    Dexter Gordon

Syncopated dance music was being played in New York from around the turn of the century by local orchestras, and ragtime piano was evolving into an idiom there that would later be dubbed Harlem Stride. By 1915 a black New Orleans band, Freddie Keppard's Creole Jazz Band, was visiting New York on the vaudeville circuit and in 1917 the white Original Creole Orchestra, also from New Orleans, was booked into Reisenweber's restaurant on Columbus Circle. The ODJB, a quintet, created a sensation and the combo was the first to record the sounds of jazz, which they did soon after their New York debut.

By the beginning of the 1920s jazz was on its way to establishing itself in New York. The city had clearly become the mecca by the end of that decade and to this day remains so. The player who has become prominent on the national and international jazz scene and who has not served an apprenticeship in the Big Apple is all but impossible to find.

"It was intimidating," alto saxophonist Arthur Blythe confessed to me about a visit he made to New York from his native Los Angeles in 1968. "It was faster, swifter. People did everything faster—walked, ate, drove—everything was faster." Arthur stayed only a few weeks, waiting six years to give it another try.

"When I came back it was less intimidating," he said, but he soon discovered that he was "not abreast at all" with recent developments in the music. "You have to live in New York to feel what's happening. You might not be able to explain it totally, but you have a sense of what's going on. It's hard to convey that to someone who hasn't been to New York, but its difference is as Paris would be or as Singapore or Germany would be. As far as this music is concerned, the music that I've been playing, if artists come from New York, they're recognized throughout the world as being more professional than if someone came from St. Louis, San Diego, or even L.A."

A proper history of the role New York has played in jazz would necessarily fill a number of volumes. That history has not yet been written, although a number of biographies and critical works dealing with various aspects of that history have appeared over the years and continue to do so at an accelerating pace. We do not propose to deal here with the entirety of that history but rather only with a few representative individuals. Once again, we shall let these artists tell their own stories, for the most part.

"I didn't know of much jazz being available in the New York area except in Harlem," multi-instrumentalist Spencer Clark told me, looking back to the mid-1920s. Spencer, who played and recorded with the California Ramblers and other bands, has always specialized in the bass saxophone, of which he is one of the great masters, but he has at one time or another played all of the saxophones and clarinets, cornet, mellophone, alto and baritone horn, mandolin, four-string guitar, bass, and harp.

"You could walk into any of a dozen clubs in Harlem and hear some great jazz in 1925, '26, and '27, along in there. I didn't get to know them well enough to know from where they had come, but the guys were great to us. They let us sit in, said we should come there and make ourselves comfortable and at home. We learned a great deal with the boys up in Harlem 'cause they were all good players and they had that feeling, that wonderful feeling that we were groping for.

"I think one of the real eye-openers to the New York musicians was the arrival of the Jean Goldkette band in 1926, 'cause that was one of the greatest things that ever hit New York," Spencer enthused. "The musicians were all absolutely thunderstruck. They outnumbered the dancers in the Roseland Ballroom twenty to one. The dancers, who cared about them?! The place was lined with musicians hanging over the rail, taking this thing in, and I'm one of them. That band was one of the most thrilling, exciting experiences of my life, as I look back on it. It was just tremendous—oh, what a band! Bix [Beiderbecke] was in the band and that great big guy whose name I can never think of slapping the bass, which we'd never heard done before. Great band, exciting band, the sound was just unbelievably good, it was new, it was fresh, it just knocked you over." (The bassist was Steve Brown and some of the others in the legendary band were saxophonist Frankie Trumbauer, trombonist Bill Rank, and drummer Chauncey Moorehouse.)

The late Joe Venuti, who arguably introduced the violin to jazz, outlined for me the routine of the touring musicians of those early days.

"You play in a band back in those days, in the '20s, you know, we would work a hotel, like we would play here in Washington at the Shoreham Hotel for two weeks like with Paul Whiteman. And we would be on the air, that is, radio broadcasts, for two weeks. We would be on every night and through that medium Paul would send out agents and we'd book one-nighters. Maybe out of that two weeks we'd book a whole month of one-nighters. We'd play the vicinity around Washington, we'd play Philadelphia, we'd play Pittsburgh, we'd go down to Virginia, we'd play North Carolina, we'd even play parts of South Carolina, and then we'd play New York.

"Now, you know, one-nighters were pretty strong back in those days," Joe continued in his inimitable fashion, a twinkle in his eye, "and although we were paid pretty good for them, the work was stiff. You were traveling in buses, we'd eat hamburgers and where you can get a bite to eat, and rehearse, and play the one-nighters. It's sort of flat, you know. We'd always look for little jokes to play on one another in the band. And in the Whiteman band we were a whole gang of pranksters.

"Well, we'd probably do about six months of one-nighters and play hotels and we were laid off in New York for two weeks. We had a vacation and Lennie Hayton was assigned to do an arrangement on 'Somebody Stole My Gal,' see, so I said to him, 'Look, we always play the same solos.' I said, 'Let's try to make it different.' He said, 'How, what do you mean?' I said, 'Well, Bix has got a little organ. He takes it up in his room.' And we're staying at the Cumberland Hotel in New York City and he was on the twenty-second floor of the Cumberland and he had this little organ up there. And Lennie said, 'Well, what would be different?' I said, 'Well, let's take this little organ and throw it out the window and whatever notes we hear, start the arrangement that way.' So he said, 'Say, that's a great idea.'

"Then, with a couple of drinks, we went upstairs and it was three o'clock in the morning and we threw this little organ out the window.

And we waited and nothing happened. So Lennie says to me, 'How are we going to start the arrangement?' I says, 'Well, let's go out and buy another organ and we'll do the same thing.' And we did. The next day we bought a little organ. We waited until three o'clock that morning and we threw it out the window. And only one note, we could only hear one note, 'Boop!' And that's the way Lennie started the arrangement of 'Somebody Stole My Gal.' And if you hear the arrangement, the old Paul Whiteman arrangement, it goes, 'boop, doot, doot, doot, doo, doo, dum.' We had fun, those little pranks we'd play on one another."

Pianist Marty Napoleon grew up in Brooklyn in the 1920s and '30s. He reminisced about those days for me. "I used to wake up at eight o'clock every morning because the iceman would come around and say, 'Ice!' I would jump up and put my pants on and run outside. He would cut the ice into blocks and he would have all these little ice chips on the back of the truck. The kids, we'd all run in back of the truck and grab little pieces of ice. It was such a joy.

"I come from a musical family," Marty explained. "My father was the oldest of five brothers who were all working musicians, including my uncle, the trumpeter Phil Napoleon. My mother played guitar, my sisters were singers, my brother Teddy was a piano player, my older brother was a drummer. While the other kids were hanging out at street corners we were rehearsing. We left the windows open and played very loud and kids came from blocks around to stand outside and look in. We were like stars. We were always into music, that's all we knew. We had a beautiful childhood."

Marty went on to associations with Louis Armstrong, Charlie Barnet, Gene Krupa, and his brother Teddy, and was a regular in the 1950s on Perry Como's television show. Working vaudeville-style stage shows in 1941 with bandleader, sometime pianist, and comedian Chico Marx, Marty recalled how "Chico waved the baton and made little side jokes and when he did 'Beer Barrel Polka' he would roll an orange on the keyboard. One time he threw the orange at me and I threw it to the drummer George Wettling. The next day everybody in the band had an orange and oranges were flying all over the stage. Chico loved it. 'Leave it in,' he said, 'it's great.' "

Several years later Marty was playing an outdoor amusement park with notorious practical joker Venuti. "It was so cold we were working with overcoats and gloves and hat and earmuffs. It was freezing. Must have been two thousand kids out there and they were all warm 'cause they were dancing. Joe sent the band boy out to buy some frankfurters and he got a chair and broke it up and built a fire on the bandstand. He sat there roasting these frankfurters while we were playing. The manager came running up, he was hysterical, 'What are you doing?!'

"Joe used to bite his lip and look at you as if *you* were crazy. He said, 'What's the matter?' . . . and the manager's screaming, 'You've got a fire on the stage!' Joe said, 'So what?' And he got up and stuck his hand in his pocket and says, 'Here. How much was the chair?' "

Singer Billy Daniels, who died in 1988, came to New York in the mid-1930s and had some very striking memories of the era. The Jacksonville-born and later world-traveled vocalist whose signature song was "That Old Black Magic" assured me that he "never intended to go into the business, never thought I had that kind of a voice, but I did love to sing when I was a kid. My dad wanted me to be a lawyer and I came to New York to go to Columbia University. I dutifully registered and started working nights in Harlem as a busboy at these nightclubs.

"Well, I started to sing as a busboy, as a singing waiter, at Dickie Wells' club, the very first job I had, and I played Smalls' Paradise. I played the downtown Cotton Club in 1939—it had moved into the Latin Quarter—with Bill Robinson and Duke Ellington and I sang in the show. I sang a song called 'Don't Worry 'Bout Me.' Now mind you, I wasn't a principal, I sang while the girls danced. I didn't reach any sort of big audience until I had worked on 52nd Street. I worked the Onyx, the Famous Door, Kelly's Stable, Manny's Chicken Koop, and then I worked Park Avenue.

"I went from Park Avenue at 52nd Street to Cafe Society and then to the Riviera in Jersey and then I wound up in the Copa. I worked Nick's and the Blue Angel. I worked with Charlie Shavers, the Clarence Profit Trio, Coleman Hawkins. Matter of fact, when Bean arrived (he had been in Europe for a long time) that night he played 'Body and Soul' and I shall *never* forget it. And of course there was Nat Jaffe, a fine pianist, Stuff Smith, and the King Cole Trio.

"While in Atlantic City in 1942 I found 'Black Magic' and of course that song went in and out for about a year or so and then I recorded it and it was an immediate hit and it took me all over the world. I've had ten Royal Command performances in different countries and I've been invited to the White House five times.

"The only thing we served at Dickie Wells' was fried chicken and biscuits, but there was no kitchen. I'd go through the back door where there was an alley leading to the next street and Tillie's Chicken Shack, and of course Tillie's chicken was very famous in Harlem. A lot of the moochers caught on to the fact that chicken was going through the alley and they tried to take it away from you. I had to hustle and of course I had someone to watch out for me. If I lost a plate they would charge me a dollar for it so I made up my mind not to let that happen. When they tried to take it away from me and knocked it off the plate I'd pick it up, brush it off and put it back on the plate. And I always wondered why the people, after I did that and asked, 'How was it?', they'd say, 'It was the best bird I ever had.'

"It wasn't considered a big deal," said Billy of the easily available marijuana, "you could buy it all over Harlem. We didn't have heavy drugs, there was no talk of cocaine, and if there was anybody we knew that was on heroin, we just ostracized him. Pot wasn't a problem at all and I don't remember anyone getting arrested except Mezz Mezzrow and he was making packages of it. We didn't have problems with it, we had fun with it.

"We had a lot of goulash parties and gumbo parties. Every place that had music would have a pot so that the musicians could eat and that made it easier for us to hang out and eat all night. What you had to do when you got to Harlem was ask, 'Where's the band, where're they playin?' It would change from one place to another. When you went out in those days, if you went to a show downtown, you'd say, 'Let's go up to Harlem,' see another show, which could be a girlie show or something. We had Noel Coward, Cole Porter, Ira Gershwin, Tallulah Bankhead, Douglas Fairbanks, Jr., Barbara Stanwyck [in the audience].

"The Mob was always present," said Billy of the 1930s New York club scene, "because originally they had the booze and when prohibition ended they had the places, you see. Nothing really changed except the laws. The places where we played usually had Mob fingers somewhere. We got to know them because they were in and out. I never had any truck with it but it was quite obvious what was going on. As a matter of fact, right after Prohibition the best beer that was put out was Owney Madden's beer and it sold in the finest places.

"I remember Red Murray, a torpedo for the Mob. A torpedo is a guy who collects and a guy who's a heavy dude, very good at breaking legs, elbows and whatnot. Some of Murder, Inc., used to come in places where I worked, but I didn't know that's who they were. They used to come in and have a good time, have a ball, a half-dozen of them that I got to know. They were always full of fun, usually on a kick or something. We never had any trouble, but one day I pick up the paper and there they all were. I said, 'Holy mackerel!' They were all grabbed at once. They had one of the leaders and he was talkin'. One of them always carried a satchel and I thought he was a salesman. I realized afterwards that he had the tricks of his trade in that bag."

Pianist Dick Wellstood, who died in 1987, was still in his teens in the mid-1940s when he made a practice of attending the performances of the great stride piano players who worked in the clubs on 52nd Street and in Greenwich Village and in Harlem bars.

"In those days there was still a lot of those guys playing around," Dick told me. "I mean, it wasn't a great big archaeological trip because James P. Johnson and Willie The Lion Smith and others, such as The Beetle and Donald Lambert, they were all playing around and all you had to do was go hear them. So if you liked their music, it didn't take a lot of work. Like nowadays it's work, you have to really dig up the records and sort of wonder what it was like.

"I used to go up to Harlem. There was a place called The Holly-wood and Monday night was piano night and a lot of piano players used to go up there, especially Art Tatum. They used to kind of stage 'battles' up there. I went up there to hear an old stride piano player named Willie Gant and they always used to have a 'battle' between him and Marlowe Morris and Marlowe always won. And I always used to get angry, but I couldn't say anything. I couldn't understand why they would let Marlowe win. Of course, Marlowe was a terrific piano player, probably a lot better than Willie Gant, but I went up there to hear Willie, see.

"There was another one named Gimpy Irvis, who was the brother of Charlie Irvis, the trombone player," Dick added, "and the reason he was called Gimpy was because he was crippled in his right leg. And he used to use this for an act. He used to take off his left shoe and then he'd play the stride bass using the big toe on his left foot."

Thus did Dick Wellstood learn the craft directly from some of those originals mentioned above, and for many years, Dick was virtually our only lifeline to the likes of James P., The Lion, and other legendary stride piano "professors." Dick never saw Fats Waller perform in per-son, for that keyboard genius died in 1943 when Dick had not yet ventured onto the club and bar circuit. "But I played every night for seven years with Gene Sedric," he pointed out, "who was Fats' clarinet and tenor player and I played a great deal with Herman Autrey, the trumpet player on most of Fats' records. Matter of fact, I recorded on Herman's last record and Gene's last record, too."

Another who learned directly from a master was the saxophone and clarinet player Bob Wilber, whose concert program "From the Cotton Club to Carnegie Hall" was launched soon after the release of the 1984 film *The Cotton Club,* for which Bob was musical director. Wilber has become a leading figure in the jazz repertory movement, which we shall discuss in the chapter on the contemporary scene.

"Basically, I grew up as a kid in the big band era,", Bob told me in one of several interviews, adding that he was born in New York City in 1928 and grew up from the age of seven in Scarsdale. "I was caught up in the excitement that all young people in the country seem to have been caught up in then. It was a very exciting period of popular music—I think we can all agree on that—and the standards of popu-lar music were very high.

"It was not too hard a transition to go from being crazy about big band music to becoming familiar with different soloists in the bands and sort of getting more of a sophisticated idea of what jazz was about. So by the time I was thirteen or fourteen I was really ripe to get into jazz and I began to find young people who had record collections of the old Armstrongs and Beiderbeckes and all those things."

It was during his teen years, the 1940s, that Bob came into contact with these youngsters of like tastes. He recalled that there were about twenty Westchester County jazz enthusiasts of high school age who hung out together. Several of them were committed to becoming jazz musicians. These included pianist Dick Wellstood, trumpeter Johnny Glasel, trombonist Eddie Hubble, and drummer Eddie Phyfe, familiar enough names to the student of this music. They traveled to New York on the commuter trains on school nights and on weekends to check out the varied music offered in the city's jazz clubs.

Wilber and his companions sought out the antiquarian New Orleans style sounds of Bunk Johnson at the Stuyvesant Casino, and they hung out Saturday afternoons at Milt Gabler's Commodore Record Shop, where they often encountered some of their musical idols. They were regulars at Nick's, where Muggsy Spanier, Pee Wee Russell, and other Chicago Style legends were the bill, and they never let pass a week without visiting Eddie Condon's, the eponymously named club whose proprietor was sometimes found on the bandstand with his guitar, along with Wild Bill Davison, Edmond Hall and Georg Brunis, to name only several of the many who made up the house band over the years. Eventually Wilber and his gang began to sit in at Sunday afternoon jam sessions at Jimmy Ryan's, and in 1947, under the name Wilber's Wildcats, they recorded on Gabler's Commodore label and for Rampart Records.

"This was an era when jazz was split into two very diverse factions," Bob observed, citing historical fact as well as speaking from personal recollection of the period. "This was a new phenomenon in jazz which didn't happen before then. There was the New Orleans Revival, when fans went back and discovered older players like Bunk Johnson, and there was the reissue of early jazz records. This coincided with the new movement in jazz called bebop. These two movements were sort of going in opposite directions and it seemed like all the fans were either going one way or the other and there was no middle ground.

"Well, a lot of musicians, including myself, were interested in creative players whether they were playing traditional jazz or the modern jazz of the day. It was all music and we were interested in the creative aspects of it. We used to hang around as kids on 52nd Street in the '40s and we were listening to Sidney Bechet, who would be playing at Jimmy Ryan's, and then we'd go across the street and listen to Charlie Parker and Dizzy Gillespie, who were over at the Deuces. I was interested in Lennie Tristano's innovations. He was a very creative, unusual musician and I got a lot out of being able to study with him. I didn't want to get typed as strictly a traditional player."

Bob cited Benny Goodman and Louis Armstrong as early influences. In fact, Bob directed and starred in the sold-out 50th anniversary Carnegie Hall tribute to Goodman, in January 1988. But if ever one artist had a *major* influence on another, it was the great Sidney Bechet's influence on Bob Wilber.

"Sidney Bechet was making records under the name of the New

Orleans Feet Warmers and my interest in him grew and grew. Through a mutual friend, the clarinetist Mezz Mezzrow, I was able to meet Sidney Bechet. He was opening a school of music at his home in Brooklyn. His idea of opening a school of music was he got a sign painted that said 'Sidney Bechet School of Music' and tacked it up outside the door and sort of waited for something to happen. Well, naturally, not much happened except Mezzrow sent me over there and I started studying with Sidney. Maybe there was, at most, two or three other students and that was really the extent of the music school. I ended up sort of being his favorite pupil, I guess, and after a couple of months he said, 'Well, gee, you're living over there in the Village and it's a long ride in the subway—why don't you move in? I've got an extra room.' So I did. I would have been eighteen when I started studying with Sidney and I lived there at the house with him and we would practice with the two horns and study and fool around all day and at night we'd hop on the subway and go over to 52nd Street where he was playing at Jimmy Ryan's and I'd play on the stand with him. So I had some great on-the-job training that I'll never forget.

"Sidney Bechet was the first great saxophonist in jazz and he was a tremendous influence on every saxophone player in that era. He was a phenomenal virtuoso, had total mastery of the instrument. When I first took up the soprano in the '40s he was *the* soprano player and that was about it. I think the soprano has been enjoying a resurgence since Coltrane took it up and became a major influence. Now every tenor player owns a soprano, but they don't all play it well because they tend to think of it as a little tiny instrument after the tenor and they get a small sound, whereas Sidney got a big sound on the soprano.

"I was listening to some records Sidney made in France in the last ten years of his life that Kenny Davern loaned me that I hadn't heard before, and I had the feeling that if Bechet were back with us and on the stand with some of these young turks of the soprano saxophone like Wayne Shorter and people like that, he would wipe them out, he would be brutal, because he got this tremendous sound, dramatic quality, drive—he had it all together. The tragic thing about Bechet is that his name is just not known in this country. Duke Ellington absolutely worshipped him and of course Johnny Hodges worshipped Bechet also."

Yet another of those precocious youngsters of Westchester County, drummer Eddie Phyfe was out on the road as a seventeen-year-old in 1946 with clarinetist Pee Wee Russell. "All I can remember from Pee Wee is, 'Shut up and drink your spinach,'" Eddie told me, chuckling over the memory. At around the same time, he recalled, he was hired for a gig at Eddie Condon's. Eddie recounted how Condon, an incurable wiseguy with a wit as dry as a martini, concluded the verbal contract with, "Oh, by the way, Phyfe, wear shoes."

"Of course, I fell behind at Mamaroneck High School," confessed Eddie, "but I went to the senior prom and though I didn't graduate, I did graduate to Claude Thornhill's band, which played the prom." A young drummer not just of promise but of proven ability, Eddie Phyfe soon found himself regularly in the company of the greats when he was still in his teens in the 1940s. Eddie's scrapbook teems with photos of himself on the stage of Town Hall with drummer Baby Dodds and reed players Bechet and Mezzrow, in conversation with Billie Holiday, in the studio with drummer Gene Krupa, playing behind Lester Young in a club. He remembered a '50s gig at Boston's Storyville, a club founded by jazz festival promoter George Wein, where "our band reserved a table up front to hear the intermission pianist, who was Art Tatum."

One of Eddie's most cherished memories is of the night that trombonist Jack Teagarden picked him up in a limo for Louis Armstrong's fiftieth birthday party at pianist Joe Bushkin's East Side duplex. "Pops played, of course, and I think Tallulah Bankhead sang." Then there was the time he dropped by a New York club to check out his look-alike, drummer Buddy Rich. "He saw me coming in the door and said, 'Hey, cover for me, I gotta do a radio show,' and he never showed up again that night." There was the morning Eddie was due at a recording session and couldn't get a cab. "A guy with a fish truck had delivered the fish and came back to see if I was still waiting. He delivered me into the hands of [impresario and record producer] John Hammond on a corner in Brooklyn with smelly drums."

Eddie Phyfe has demonstrated over the years a facility in all of the major styles of jazz short of the free form that came into being in the 1960s. "Big Sid Catlett showed that to me. Think about it, he played with Louie, he played with Prez, he played with Bird," Eddie points out, citing Armstrong, Lester Young, and Charlie Parker as representatives of the spectrum covered by Catlett, whom some believe to have been the greatest drummer in jazz. "He was a guy who could play anything, big and small, and always musically. He was my big idol, along with Dave Tough. That's the kind of drummer I was trying to be."

Eddie's memories echo those of his erstwhile fellow bandsman Bob Wilber. "There were labels being attached to everybody during that era," Eddie hastened to point out, "you know, the 'moldy figs' and the 'sour grapes' and all that nonsense, but they meant absolutely nothing to me. I just played music and most of the people I listened to played anything. The avant-garde of today is not any more far out than what we were doing back then in sessions, which was also shocking at first. But you got over that pretty quick.

"They had all kinds of things going on on the same street," Eddie said, providing illustrations from his own professional experience of keeping the beat going in the course of one evening for stride player

Willie The Lion Smith and bopper Bud Powell. "Georg Brunis took
his [traditional style] band and marched across the street playing 'The
Saints Go Marching In' to the club where Dizzy was playing one night.
Then Dizzy got seventeen guys and marched over to Jimmy Ryan's
and played 'Salt Peanuts' or something in the door."

Fifty-second Street between Fifth and Sixth avenues had by the late
1930s become known as The Street and it remained for a decade the
jazz center of the world.

"My god, you couldn't walk without bumping into all these luminaries," exclaimed Mason "Country" Thomas to me in one of several
interviews I conducted with this Washington, D.C.–based multi-reed
player. "Country" was thinking back to the 52nd Street he visited in
1943 just before he went into the army. He had made friends as a
teenager on overnight New York visits with clarinetist Mezz Mezzrow,
guitarist Eddie Condon, clarinetist Pee Wee Russell, and saxophonist
Coleman Hawkins.

"Those were some very legendary days," vibraphonist Milt Jackson
told me, looking back with awe to 1945 when he was brought from
Detroit to New York by Dizzy Gillespie. "It's just amazing, man, when
I think back and go over the lists of artists who have since passed on.
All of the artists that I had been reading about in magazines and
whatnot, like in *down beat,* for example, I got a chance to see because
just about every known musician in the business was in New York.

"Can you imagine, for example—something that doesn't exist today—Billy Eckstine being in one nightclub and about two doors away
Sarah Vaughan would be appearing in another, Billie Holiday would
be appearing across the street, and probably in a club three or four
doors up the street from there Dizzy Gillespie's got a big band, Coleman Hawkins had a quintet? Just about everybody you could name,
anybody you ever dreamed of hearing or seeing in those days, was
right there, yeah. Art Tatum, Lester Young, Charlie Parker, Hot Lips
Page, Fats Navarro, you know, every artist that was in the business
and of any note whatsoever, man, I got a chance to see and hear,
more or less, at one place or another."

Tenor saxophonist Big Nick Nicholas was another who arrived in
the city at about that time, coming from Lansing, Michigan, by way
of Boston. "When I came to New York they had all these famous
people that I'd read about and listened to their records," he related,
adding that the clubs on 52nd Street where he hung out "were like
two or three feet from each other. You could hear Art Tatum here
and go across the street and hear Billie Holiday or Erroll Garner or
whoever there.

"Up the street was my idol, Coleman Hawkins. I used to dress like
him—of course I couldn't afford those expensive clothes he was wearing, but I had reasonable facsimiles—and my moustache was like his.
I kept coming there every night with my horn. All the bartenders and

all the waitresses knew me and they'd say, 'Here comes Coleman, Jr.'
I *loved* that! One night I came in and sat down and he says, 'Well, son,
you feel like blowing one?' He said, 'I think I'll play 'Body and Soul,'
which was his masterpiece. And so I says, 'Yes, sir!' So I went up and
played and he just smiled. From then on I was *in*, because I played
just like him and when he wanted to go someplace, he'd take off and
come back an hour later and I'd be Coleman Hawkins for an hour."

The drummer Roy Battle told me that he "grew up in Brooklyn
and used to work down on 52nd Street. I stayed on The Street for a
long time. I was the house drummer at the Three Deuces and I played
at the Spotlite. Each club would have two bands and you alternated—
each band would be an hour on and an hour off. There was a saloon,
a stand-up saloon called Pat's, on the corner and we would all hang
out there. Somebody would come and say, 'Show time, Three Deuces!'
and all the musicians would run. We would get through about four-
thirty in the morning and then we'd go and jam at Minton's. I'd come
home maybe six or seven o'clock and we'd stop the milkman and drink
the quart of milk right there. Or we would all go to Bickford's, all the
show people would meet at Bickford's, and we'd get breakfast there
and just talk.

"Big Sid Catlett took me under his wing and showed me how to use
my brushes so that I could play for Billie Holiday. He was so smooth.
I worshipped him. I recall the time when Billie would come on. Like,
the tables wouldn't be over, I guess, like sixteen inches around, you
know, and they would turn off all the air conditioning, they wouldn't
serve anything, the waitresses wouldn't take any orders, everything
came to a standstill. And she would come on and just be Billie. It was
just that she was the queen. It was such a beautiful thing.

"During the break we'd go upstairs where the musicians were and
she would have an enclosed place where she could keep all her gowns
—she had about a hundred gowns and a hundred pairs of shoes—
and she had a dog called 'Mister.' One night she was up there crying and I asked what was wrong and she said that 'Mister' had chewed
up about ten pair of her best shoes."

Of the many giants alluded to already in this chapter there is one
whose name keeps coming up, the piano virtuoso Art Tatum. Billy
Taylor spoke to me about his friendship with Tatum. "The first time
I met him was when I auditioned for a job with Ben Webster at the
Three Deuces in 1944 and Tatum's was the other group there. It was
the beginning of what was for me a most important friendship. He
was just a marvelous friend. I didn't realize at the time that he would
tell people from time to time, 'Hey, there's this kid that plays piano
and he's pretty good, you ought to give him a job.' So I was getting
some work and I thought, 'Gee, people are recognizing me as a piano
player.' And he never told me. I found this out from other people."

Billy would go along with the nearly blind keyboard genius to after-hours clubs. "When he finished work he liked to go and hang out a little bit and listen to other players and singers and just drink beer and have some fun. There were some sessions that would take place at a place called the Hollywood Bar in Harlem. Every Monday piano players would start gathering about nine or ten o'clock. Tatum wouldn't show up until maybe about one or two in the morning and by the time he got there all the heavies would be shootin' their best shots. So after a while, after everybody had played, Tatum would get up and play and give everybody a master class on how it should be done."

The bassist Slam Stewart, who died in 1987, was a member of the Art Tatum Trio for several years in the early '40s. The seeming complexity of the recordings that the trio left us can easily leave one with the conviction that the numbers were planned out beforehand and carefully rehearsed. I asked Slam how much preparation was done before they mounted the bandstand or entered a studio.

"The answer is very simple," Slam responded. "As a matter of fact, the Art Tatum Trio, when we got together, along with Tiny Grimes on guitar, we didn't rehearse very much. We actually made things up as we went along. Say, for instance, we were playing an engagement, we were learning as we went along. And Art—the piano that man played! What I mean, it was fascinating and, of course, I imagine that both Tiny and myself, we were blessed with the capacity of being good listeners and keeping our ears open to what Art Tatum was playing at the time. And that *is* a blessing when you can hear what's going on and you can sort of blend in what is being played, that communication. It was quite an experience, I might say, playing with Art Tatum, he was so great."

Art Tatum was an important transition figure and, in fact, his impact upon Charlie Parker was considerable. Tatum came out of stride piano and he acknowledged his debt to Fats Waller with pride. His harmonic sophistication was advanced and his technical skills were awesome. These two elements of his art made him especially appealing to the young generation of jazz musicians coming up in the 1940s who were putting together the basics of modern jazz, the style that succeeded swing and which soon came to be known as bebop.

Another transitional figure was the tenor saxophonist Dexter Gordon, who grew up in California and went on the road with the Lionel Hampton band in 1940 at the age of seventeen. In several wide-ranging conversations during the first couple of years after he had returned from a decade and a half of expatriation in Europe, Dexter provided me with an insider's view of these transitional years.

"It was fantastic," Dexter began, echoing the sentiments of several we have already heard from, "because The Street, in this one block, in fact, in the radius of a half a block, there must have been seven, eight, I don't know how many jazz clubs and everybody you could think of worked on The Street. You know, Billie Holiday, Art Tatum,

Dizzy and Bird, Sid Catlett, everybody. When I left the Billy Eckstine band in '45 I stayed in New York and started working on The Street and I worked with Bird. We had a sextet, Bird and I, Miles and Stan Levey and, yeah, Bud Powell and Curly Russell. That went on for a couple of months and then I had my own group, quartet style, and I worked at the Three Deuces, the Spotlite, the Down Beat, Onyx. It was a really unbelievable period in time. Fifty-second Street, it was something."

The conversation went on to consider, among other things, the principal influences upon Dexter, whom I knew to be Lester Young and Charlie Parker, and he responded in both general and specific terms.

"Lester's contribution was so great, I mean, it's hard to really answer that. In the middle '30s, when he started recording with Basie and so forth, he had a seemingly very different—it *was* very different—and new concept of playing, not just saxophone but music, what he was relating. And he had such a *joi de vivre,* so much love of life in his playing, you know, that was very, very infectious. I think everybody that heard it was—'Ah-h-h,' they'd say, 'Oh, boy!' Because what he was doing was in direct contrast to what Coleman Hawkins was doing, and Chu Berry and people like that were doing. Like he was dancing with his music, with his horn, and musically speaking, his harmonic approach was very colorful and tasty because he was really the first one to start playing ninth chords and sixth chords and making use of these color tones.

"Up to that time most of the swing musicians were playing pretty basic harmonies and Lester was the first, one of the first anyway, to use these color tones. To give you an example, Debussy writes using these types of chords that are very pastel. And it seemed like with Lester his solos were always so well constructed and they always told a story. In fact, the cats on the corner at the lunch counter around the juke boxes, they could all sing Lester's solos and they had words for them. To me, that was where it was at, the real approach. He also told me that knowing the lyrics, understanding the lyrics to ballads and things you're playing, it helps you to understand the composition, which I have found to be quite true.

"Yeah, well, the Bird was all they say—he was really electrifying. Bird was from Kansas City so he knew Lester and grew up listening to Lester and the Basie band, Benny Moten, Hot Lips Page, and people like that, so that's where Bird was coming from. I mean, you can hear a lot of Lester in Bird, but what Bird did, now, was to extend what Lester had been doing. So that's where we started getting into extending the chords to the eleventh and thirteenth and the augmented elevenths, the flat fives, and things like this. Like I say, up to that time the harmonic thing was limited and when Bird came he extended the harmonic thing, playing changes, playing chords on top of chords that would relate to each other. And of course Dizzy was doing the same thing. It was just evolution, it was just electrifying. And it was amazing because he was just on top of everything, all the new tunes, the show tunes, and seemingly he knew just everything. Somebody would say, 'Wow! Who's the—where'd *he* come from?!'

"I was very much influenced by Lester. I don't mean that I was trying to play note for note what he played, but his ideas, his philosophy and approach to music and to the tenor is what intrigued me. And then Bird came, like I say, which was an extension of what Lester and others had been doing. Fortunately, my timing was just perfect because I came up in the early '40s and one of the first cats I met when I came to New York *was* Bird. He was with Jay McShann's band. And through going to Minton's where all the cats used to go to jam—which was a school in itself, you know, playing with Thelonious and Dizzy, Kenny Clarke, Bud [Powell], everybody—I just grew up in that era and so what came out of that was Dexter, huh? And like I say, I think that by more or less the same process, when Trane came up and Sonny Rollins and Jackie McLean, which were kids when I first went to New York and was on the scene, . . . I guess they heard what I was doing and felt that was valid and then worked on their own thing from what we all had been doing. Like I say, I think this is the evolution of the modern tenor. And that's the beautiful thing about jazz, that it's always changing."

Dizzy Gillespie told me, "It's a shame that Minton's is boarded up. There were two places, Minton's and the Uptown House. They were on an equal basis. The creativity in both of those places was fantastic. You could learn a whole lot by just sitting in because all the guys from all parts of the country and all parts of the world used to converge on New York and come to Minton's and after hours go to the Uptown House. Don Byas, Lester Young, Coleman Hawkins, Big Sid Catlett, Hal West, Kansas Fields, Jack Parker, Nick Fenton, Monk, John Lewis. By the time John Lewis and those came on the scene in the late '40s the music was developed to a point where you could really see that it was going some place and all of the new guys—Ray Brown, Clifford Brown, Miles Davis, Fats Navarro, Kenny Dorham—all of them had specific voices but all of it was based on what we had done. And then it went down on 52nd Street, where the money was, and it jumped up to Broadway at the Royal Roost and Bop City and then, finally, at Birdland.

"I was speaking to Thad Jones recently. He was in the army down in Hawaii in 1945, one of those islands out there, and he and this baritone player heard 'Shaw Nuff' for the first time. He said he fell out of his bunk and he said he laughed all night, his jaws were sore from laughing so much, and he said, 'It finally is here,' he told the guy, and they were followers after that."

A query I put to Dizzy about his early background elicited a rather remarkable summary of his career. "I was born and raised in Cheraw, South Carolina. There was a lady named Miss Alice Wilson, a teacher in my school in Cheraw, and we had a little band. We only played in B-flat but we used to jump. We had a bass drum player, snare drum player, a piano, trombone, and that was our band. We played for the kids marchin' and we had a minstrel show every year. That was beautiful, it was just nice, and that developed and I learned how to read music and started composing and things like that. [Later] I made a television program in my home town and used Miss Alice Wilson as a catalyst. It was beautiful. She played the piano and we talked about old times.

"We didn't have any records in those early days. Only the people with money had records. We didn't even have a radio at first, so we had to listen to what we could. I started on the trombone because that was the only instrument that was left and I started on trumpet when I was about thirteen. I used to listen to the performance from the Savoy Ballroom, Teddy Hill's band. Roy Eldridge was the trumpeter in that band and I fell in love with his style, and thousands of other young trumpet players fell in love with his style at the same time.

"I moved to Philadelphia in 1935 and got a job right away. I played in a band with Charlie Shavers and Carl Warwick in Philadelphia called the Frankie Fairfax Band. I went back down south in the latter part of '35 or early '36, went back down to Charlotte, North Carolina, with Tiny Bradshaw, came back to Philadelphia and then Charlie Shavers and Bama—Carl Warwick—cruised me over into New York in 1937 to go with Lucky Millinder's band, which I didn't get the job. That time I was supposed to take Harry Edison's place and Lucky decided to keep him. About that time I got a job with Teddy Hill's band and went to Europe in 1937, came back and played around with Teddy Hill's band and with Cass Carr, Edgar Hayes, Claude Hopkins, Fess Williams, Calvin Jackson, Mercer Ellington—Mercer had a band when his father was alive. Then I went with Cab Calloway in 1939, left there and went with Ella Fitzgerald, Coleman Hawkins, Charlie Barnet, Benny Carter, and Lucky Millinder; left Lucky in '42 and went with Earl Hines along with Billy Eckstine, Sarah Vaughan, Charlie Parker; left Earl Hines in '43 to be the musical director of Billy Eckstine's band in '43; stayed with Billy in 1943 and '44; '44 I went into the Onyx Club as the co-leader of a band with Oscar Pettiford. That broke up in a while and I was co-leader with Budd Johnson across the street on 52nd Street, and then I went into the Three Deuces with Charlie Parker and Max Roach and Ray Brown. I played with Duke Ellington, Les Hite, all the bands around New York, and my first big band was in 1945. The historical [small] band was the band that I took to California in 1945, latter part of '45, with Milt Jackson, Charlie Parker, Al Haig, Stan Levey. I came back to New York in early '46 and organized another big band at the Spotlite. I had Thelonious Monk, Kenny

Clarke, Al McKibbon, and then the pianist changed to John Lewis, and then Chano Pozo came on the scene and in '48 I took that band to Europe.

"It was only a matter of evolution," Dizzy summed up. "That is the story of our music. It evolves and each age has its heroes. So I was the hero of the '40s, then somebody else would come for the '50s, another one for the '60s. But the music of the '40s, the music of Charlie Parker and me, laid a foundation for all of the music that is being played now. So I feel very fine about that, you know, that you've been an influence for musicians to go further."

The account of each city or region that we have so far discussed—and of those yet to be discussed in this survey—could easily be brought up to the present and in every case is a history that must be written if we are one day to achieve anything approaching an adequate documentation of this art form. We have attempted to zero in on the peak years of the style or styles that define each of the locations or areas and have chosen from our sources with that in mind. In the case of New York that approach is not especially practicable for a strong case can be made that the *past six decades* constitute New York's peak years. Even a California-reared artist whose roots are in the Swing Era, and who resided abroad for fifteen years in the 1960s and '70s, Dexter Gordon, could state with total conviction in 1977 that "New York is *it*, has always been, it's the mecca, the testing ground, and I can't see any other change."

And so we leave to succeeding chapters the subsequent developments that called New York home, having confined our New York chapter to the 1940s scene of Harlem, Minton's, 52nd Street, and such landmarks on the musicscape of the decade as the New Orleans Revival.

5

The
Big
Bands

Instead of the hi-fi set, people played music.
Jaki Byard

These bands were my college, my university training.
Dexter Gordon

You reach a certain place in your development and if an audience won't let you express yourself as you must, anything else you do is fraudulent.
Artie Shaw

The Big Band Era has traditionally been assigned to the years 1935 to 1945, although some jazz historians prefer a less precise bracketing and refer to the period as running from the mid-1930s to the mid-1940s. We do not quarrel with the dates as representing the decade or so during which the big bands, the music they produced, and the vocalists and instrumentalists they featured constituted *the* popular musical form of the time. But we hasten to point out that large musical units employing jazz expression had been around for some years. The genesis of such units can be dated to the early years of the century when brass bands, most including some reed instruments, flourished. The definitive account of big-band jazz activity (which mammoth task has yet to be undertaken) will clearly have to explore those beginnings. While we do not propose to do so here, we shall offer several pre–Big Band Era scenarios as background to the decade alluded to.

"I went to a school in 1918 in Manlius, New York, called the Manlius Military School and I suppose that's where I got *real* interested in music," trombonist Spiegle Willcox recalled for me at the Manassas Jazz Festival in Virginia. "There was a brass band there at the time that got into the jazz thing and I played soon after that with Cornell bands. That was in early '21, '22. And there was a band called the Big Four out of Cornell and they were finally known as Paul Whiteman's Collegians. That lasted to [an engagement at] the Chase Hotel in 1924.

"Then in 1925 I played all summer in Auburn, New York, at a park, and Fuzzy Farrar, who was the lead trumpet player with the Jean Goldkette band, happened to be at the park at the right time and came up one day and, you know, sat around with the band and said, 'I'm here on a vacation.' And I said, 'Where you from?' and he said, 'Detroit.' He said, 'Yeah, I play with a band out there,' and I said, 'Who do you play with?' He said, 'Jean Goldkette,' and I said, 'Jean Goldkette! What are you doing tonight? Can you come and sit in?' He said, 'Yeah, I'll come over.' And he came over and sat in and, to make a long story short, he said, 'How'd you like to join Jean Goldkette? Tommy Dorsey's leaving.' 'Well,' I said, 'I have a commitment with the California Ramblers. I've got to fulfill that for the next two weeks.'

"This is Labor Day, 1925," the seventy-seven-year-old trombone player continued. "So I went down to New York and, if my memory will serve me right, I just didn't like New York—it was too busy. So I had three wires from Charlie Horvath, the manager for Goldkette, and I'm thinking, 'I'se going, I'm going to Detroit.'

"And I went off there and I was immediately put in at the Graystone Ballroom—it's torn down now, you know. When I went down to the Graystone that first night the only guy I knew was Fuzzy. Who was fronting the band? Russ Morgan. In the sax section was Fud Livingston, Doc Ryker, and Don Murray. In the trumpet section was Ray Lodwig and Fuzzy—no Bix yet, see. And my sidekick was Bill Rank. Steve Brown on bass, Chauncey Morehouse on drums. And Paul Mertz—he was one of the piano players.

"After a couple of weeks in come a guy by the name of Jimmy Dorsey, and he lasted six months. Russ Morgan was getting to be a great success as an arranger so he left us. We had a new front man.

"Then in March of 1926—I must say I had my bride with me, fifty-five years coming up December 14th. How about that, for a musician? Pretty good? In March Jimmy left us and in came Bix Beiderbecke—or Frank Trumbauer *with* Bix. And then right away we had three horns, three trumpets, a cornet and two trumpets, and it began to shape up. We also had a fine arranger in Bill Challis. Oh, God bless

Bill. So we were whippin' up some pretty fine stuff, and every now and then we would go to New York and play in the Roseland, for keep, I suppose, and also to make a Victor record."

Spiegle Willcox left the Goldkette band in 1927 and settled into a family business, continuing to lead his own band in the Syracuse area for years. He had been inactive musically for a spell when in 1975 he was invited to participate in a Carnegie Hall tribute to Bix Beiderbecke. Spiegle was one of a mere handful of surviving alumni of the Goldkette band to take part in that celebration and to appear a few months later in a Goldkette reunion at the Newport Jazz Festival in New York. Among those featured in the two events were Spiegle's fellow trombonist Bill Rank, drummer Chauncey Morehouse, and violinist Joe Venuti, whose friendship Spiegle had enjoyed for half a century.

"It was Joe Venuti," Spiegle insisted in explanation of his return to playing. "It's why I'm here today." Venuti had invited Spiegle to join him on gigs here and abroad and the trombonist took him up on it. Spiegle was astonished upon observing the reception given Venuti in Italy. "The Italians loved him," he said.

Joe Venuti's passion for practical jokes and his bizarre sense of humor had defined his personality as far back as any of his many acquaintances can remember. A collection of the anecdotes told about him would fill a volume. "We'd be here for five hours if I got to talking about Joe," Spiegle said, chuckling. Yet he did offer one story in illustration of the violinist's impish ways.

Venuti invited Spiegle to join him on a week-long engagement in the late 1970s at Blues Alley in Washington, D.C. Because he had another commitment Spiegle made it clear to Venuti that he could not be present for the first three days.

When the club management inquired why Spiegle wasn't there on opening night Venuti told them, "Oh, Spiegle had a fire up home and he's settlin' up a lot of stuff."

"And I got down there three days later and a guy says, 'How bad was your fire?' I said, 'I didn't have any fire.' I said to Joe, 'Joe, how come you told them I had a fire?' 'Oh,' he said, 'I had to tell 'em somethin'.'"

At around the same time that Bix Beiderbecke and Frank Trumbauer joined the Goldkette band and Trumbauer took over as front man, the Savoy Ballroom opened, in March 1926, on Harlem's Lenox Avenue between 144th and 141st streets. It would remain the principal testing ground for jazz big bands, black and white, for years to come.

"In 1929 I was in George Washington High School and we had a pretty good band," singer and dancer Peter Dean told me on my radio show over WPFW-FM in 1984. "We had Van Alexander, who wrote 'A-Tisket A-Tasket' for Ella, Butch Stone, who was with Les Brown,

Helen Ward, who was the first girl singer with Benny Goodman, and a few other celebrities.

"Well, I started this club at the Savoy Ballroom every Friday night because we didn't have school the next day. We went up to the Savoy Ballroom and plunked down our thirty-five cents and had a plate of spaghetti for a quarter or something and we would hear the most *incredible* music in the world. There were two bands, there were two bandstands. Every evening there were two bands and when one band would finish the other band would be getting in its place to play all that great jazz.

"And talk about the Savoy Sultans," Peter enthused about the house band, "They had this drummer who I'd try to copy all the time and emulate that guy, 'cause he would make wonderful movements, all flashy movements, and drum like mad. They had *some* band. It was a very exciting band. We used to run into bands like Chick Webb, Vernon Andrade, Teddy Hill, and Willie Bryant. I even heard a battle of music with Chick Webb and Benny Goodman, and by god almighty, Chick Webb cut Benny's band! Chick was one of the great drummers of all time and of course Benny had Gene Krupa, so it was quite a night. I think at the Savoy Ballroom there were very few bands that could cut Chick Webb because he knew how to play that room, and boy, did he play! And he had great men with him, of course.

"There was nothing like that ballroom. It was a block long and the floor was a beautiful parquet all varnished and shellacked so that everybody could dance. And I learned to dance. I'd grab one of those 'high yellahs,' as we called them in those days, gorgeous girls, and I learned how to lindy hop. On the backbeat that floor would go up and down"—Peter snapped his fingers to the recaptured rhythm— "and it was the most marvelous—well, it was really pandemonium, pardon the expression, but it was. I just *loved* it and everybody else loved it."

Peter "Snake Hips" Dean went on to become the personal manager of Paul Whiteman as well as other artists and he continued to perform, often accompanying himself on ukulele, into the 1980s. His vivid recollections of the Savoy ballroom in the late 1920s through the 1930s give a sense of the excitement generated by big band jazz in its heyday. He delighted so in describing the scene that I could not help but conclude that virtually all subsequent experiences paled in the light of it. Peter died in 1988.

"That is what jazz is all about," is the unqualified view of Panama Francis, who was drummer for the Jimmie Lunceford band in the 1940s and who in the late 1970s formed his own Savoy Sultans, a scaled-down version of the original unit. "Jazz was a music that people danced to as well as listened. It wasn't a thing that you sit down and listen to like at a concert of a symphony or something. That's why, if I am playing a concert, the people are sitting there patting their feet,

they're shaking their heads, I mean, they want to get up and dance!"
Panama knows whereof he speaks, for not only has he these recent
experiences from which to assess the terpsichorean inclinations of his
audiences, let it be noted that he also played the Savoy with the
Lunceford band, no doubt on some of those evenings when the late
Peter Dean was in attendance with his high school chums.

If any band comes to mind as symbolizing, if not defining, the Swing
Era, it is the Benny Goodman band. We do not in any way minimize
the seminal contributions of the many great black bands that had come
up in the 1920s and continued to hold audiences in awe throughout
the ensuing decades, several of them into the 1970s and one, the Count
Basie Orchestra, continuing to travel the world after the death of its
leader in 1984. They were, after all, the originators of this music that
we call jazz, and all of the major innovations and changes of direction
over the years have arguably been initiated by black artists.

Yet it was Goodman who brought the music of the big bands to the
attention of the general public in the mid-1930s, and it was he who
dominated the era as bandleader and soloist. However, no one could
have anticipated, from the standpoint of the early summer of 1935
(by which time Goodman had had his first band together for about a
year), that a year mostly of disappointments would follow. This is how
singer Helen Ward, who joined the band in late 1934 and would remain with it for almost exactly two years, tells it.

"We were doing a summer tour, ten thousand miles in one summer," Helen began on a hot and humid July afternoon in a back corner of a Falls Church, Virginia, restaurant, recalling with an almost
eerie command of detail the events of half a century before. "We started
across country and I guess you've read about the terrible egg we laid
in Denver—everybody was across the street listening to Kay Kyser!
Benny was very discouraged, but Willard Alexander of MCA talked
him into going on.

"We finally reached the coast and one of the first places we hit in
California was Oakland. There were mounted police on the sidewalks,
the crowds were so tremendous. Benny turned to me and said, 'My
God, we must be here on the wrong night,' because prior to that nobody liked our music. There was a picture of Guy Lombardo out front
and Benny said, 'That must be it, they're here for Guy Lombardo.'

"But lo and behold, it turned out that we were there on the right
night and these kids were there for Benny Goodman. It was our first
taste of the mob scenes that used to attend the concerts that followed.
And of course then came the famous Palomar Ballroom in Los Angeles. We were booked in there for two or three weeks, something
like that, I don't recall off-hand.

"There was wall-to-wall people—it was astounding, and needless to
say, such a thrill, just *fantastic*. And Benny, his attitude was, 'Well,
since we've had such bad luck with all our great numbers, instead of

starting out soft, let's do the killerdillers first.' Well, that's when the
crowds went crazy. Number one, the place, I would say, was about a
block square, it was *huge*. And there was just no dancing at all, the
people just stood around the bandstand, just wall-to-wall people, and
screamed. After each number we became a little more courageous
and did more of the music that Benny tried so hard to promote—
swing. The scene was just unforgetable, just an ocean of people.

"They billed Benny as the King of Swing and the amusing thing is
they called me 'America's Premier Bluestress.' If they had called me
the 'Queen' or 'Princess' or something pertaining to the King—but
'Bluestress'? This was a new one on all of us and it just went away.

"Anyway, while we were there it was fantastic and after that we
worked our way east to open at the Madhattan Room in New York—
it was then in the Pennsylvania Hotel—and on the way we were booked
into the Congress Hotel in Chicago for three weeks. We stayed six
months and every night they had ropes up. It was just fabulous. We
got to know everybody. Chicago was a little big town and everybody
knew everybody there and I just loved it.

"It had been heartbreaking on the way to the coast but on the way
back east it just seemed to grow. And of course, coming into the Big
Apple, which we called New York in those days, was another thrill,
too, the crowds and the wonderful reception. I remember there were
banquettes around the room at the Madhattan Room and the kids
would all be standing on the tops of the banquettes, not the seats.
There just wasn't room for all the people. It was remarkable.

"The thing that started it all was the NBC 'Let's Dance' program,"
Helen explained, "a five-hour broadcast, a steady five hours on the
radio, unheard of for the time. That was so the coast could get it at a
reasonable hour. We played back east until one o'clock in the morning
so they could get it out there from nine to ten. There were twenty-
five bands that auditioned for the show and Benny's was the hot band
picked. The Russ Columbo and Xavier Cugat bands were the sweet
bands." Thus it was largely the time difference between coasts that
brought the Goodman band, which routinely played late in the program, to the attention of many young people in California, causing
them to lie in wait for the band's appearance at the Palomar Ballroom.

Helen Ward had wanted from an early age to become a concert
pianist. Her father played the piano and had a gift for picking up by
ear the popular songs of the day. It was apparent that she had inher-
ited his ear, and her playing, especially when she sang along with it,
enhanced her reputation at New York's George Washington High
School as a musically talented young woman.

"The kids at school had an awful lot to do with my becoming a
singer," she confesses. "They had the bright idea that I should sing
on one of these little aluminum records, so we walked all the way

from 193rd Street to Times Square where they had one of these little shops where you could record your voice. Sure enough I went in there and I remember recording 'You Call It Madness But I Call It Love,' a Russ Columbo hit at that time. That really dates me, huh?" The song was published in 1931 and that very year had become a sensation.

Half a dozen dance orchestras had been briefly blessed with the presence of Helen Ward before she joined Goodman in 1934 at the age of eighteen. Her colleagues for the next two years included Bunny Berigan, Gene Krupa, Lionel Hampton, Teddy Wilson, and Goodman himself—pretty heavy company to keep, when you think of it. Helen left the band in December 1936 to marry Albert Marx, her first husband and the father of her only child, a daughter.

Helen Ward had been married for about a year when she and her husband attended the famous Carnegie Hall concert, which featured the Goodman band with eminent guests from the Duke Ellington and Count Basie orchestras, including the leader of the latter and tenor saxophone giant Lester Young. "Unbeknownst to me," Helen said, proudly, "[my husband] had the entire concert recorded as a gift for me, and that's how come people today can hear it." (Goodman was given a set of the discs, stored them in a closet and forgot about them for a dozen years. They were then released as an LP album and have become one of the best-selling jazz records of all time. In fact, Columbia Records has recently digitally remastered the two-record set and reissued it on compact disc.)

During the 1940s Helen toured with Harry James, did a stint with Bob Crosby and Johnny Mercer on the "Camel Caravan" radio show, and tried her hand at producing for radio. The '50s found her busy with guest appearances on radio and television and still recording. In the '60s, having settled in Chappaqua, New York, Helen was persuaded by a local dramatic society to take a leading role in a production of *Finian's Rainbow,* which she thoroughly enjoyed. "I had to learn the brogue," she laughed, then began to speak in a thick version of it, "and I found myself talking Irish even when I was offstage, I got so used to it." Helen was still singing on gigs in the 1980s, and in 1981 she recorded an album with some jazz greats, including cornetist Ruby Braff, trombonist Vic Dickenson, tenor saxophonist Al Cohn, guitarists Steve Jordan and Bucky Pizzarelli, bassists Milt Hinton and Slam Stewart, and the outstanding young Swing Era style drummer Butch Miles.

"Benny had a marvelous conception, a feeling he wanted to convey," said Helen, assessing Goodman's contribution to the art form of jazz, "and he sure did that. We remained dear friends through the years and, as a matter of fact, my husband Bill Savory and I went up to New York to see him on May 16th last year. As usual, we sat on the couch in his apartment and he looked wonderful. Right behind his shoulder was the open clarinet case where he'd been practicing. He never stopped practicing, never stopped playing. Then, by God, June 13th it was all over."

Oddly, Benny Goodman took a newly organized big band out on the road only a few months before his death in 1986, the first such large unit under his leadership for decades. One cannot help thinking that Goodman, observing the flourishing big band activity of the 1980s, wanted to have his final say vis-à-vis the idiom he helped to create as an American institution. I reviewed a February 1986 visit of the band to Washington, D.C., for the *Washington Post*.

Back to Benny's Big Band Era
Goodman's 14-Piece Unit, Stomping at the Kennedy Center

Benny Goodman's fans, young and old, filled every seat of the Kennedy Center's Concert Hall Saturday to check out the veteran jazzman's recent return to the big band format. They were not disappointed. The leader's legendary skills were there in good supply, the program of Fletcher Henderson arrangements offered example after example of the writing craft at its best, and the 14-member unit swung its socks off. And what sheer pleasure it was to experience all of this in an acoustical setting with only a single front-stage mike in use for vocals and several clarinet features.

"King Porter Stomp" and "Wrappin' It Up" had the band roaring with abandon; "Blue Room" was a model of brass and reed section precision and cooperation. The leader's clarinet was often heard soaring in high register over the full ensemble and he offered, in charming duo with guitarist James Chirillo, a wonderfully woody "Smile," which Charlie Chaplin composed for his 1936 film *Modern Times*.

The young players Goodman assembled for the occasion impressed with their ability to play strictly within style and with their excellence of technique. Outstanding soloists included trumpeter Randy Sandke with his Jamesian fire, big-toned tenor saxophonist Ted Nash, trombonist Dan Barrett, who combined the brash and the creamy smooth, and pianist Ben Aranov, whose light touch and striding momentum recalled Jess Stacy. Carrie Smith belted out "Gimme a Pig Foot and a Bottle of Beer" from Bessie Smith's final recording session in 1933, on which a 24-year-old Goodman had been present. Considerable credit for the band's energy and rhythmic flexibility goes to drummer Louie Bellson, a star himself.

The life and career of Duke Ellington has been much written about over the years although the great bandleader's immense contribution to our culture has not yet been accorded the definitive account that it so obviously deserves. The best that can be said of the several most recent biographies of Ellington is that they either present nothing new or distort the man's personality, character, and artistic ethos almost beyond recognition.

Basic to an understanding of Ellington are the several volumes of interview-based documentation that have been compiled by Stanley Dance: *The World of Duke Ellington, Duke Ellington in Person: An Intimate Memoir* (by Mercer Ellington with Dance), and *The World of Swing*. Rather than attempt a survey of Ellington's place in the music, it might be illuminating to spend a few pages with several of Ellington's vocalists. Apropos of this line of thought is a conversation I had with Dance in the mid-1980s, by which time the British-born critic/historian had been writing on jazz for five decades or so.

I pointed out to Dance that, although much attention has been directed to Ellington's utilization of the idiosyncratic talents of his instrumentalists and to his conception of the orchestra as an instrument in itself, much less attention has been devoted to his use of vocalists. This is despite the fact that the band's nearly half-century history under its founder is studded with the names of singers of singular ability, including Adelaide Hall, Ivie Anderson, Betty Roche, Joya Sherrill, Herb Jeffries, Kay Davis, Jimmy Grissom, Alice Babs, Anita Moore, and Al Hibbler. Mel Tormé performed with Ellington on several occasions, as have Ella Fitzgerald and others. Jimmy Rushing and Teresa Brewer recorded with the band and Joe Turner sang in the 1941 Ellington show "Jump for Joy." Several band members added their voices to the orchestra's efforts, notably trumpeter and violinist Ray Nance, and a vocal trio of cornetist Rex Stewart, baritone saxophonist Harry Carney, and bassist Hayes Alvis was used on occasion.

"Ellington once said to me something to the effect, 'I don't really have much of an ear for singers,' " Dance recalled for me. Dance hastened to observe that the records belie Ellington's diffidence, pointing out his use of, for example, Hall's voice as instrument, Moore's for its scat fireworks, and the "exceptional voice" of the Swedish singer Alice Babs, for whom he wrote pieces. Dance also singled out Sherrill as "a really excellent band singer, and Ellington had to play a lot of dances, one-nighters where you had to have a band singer."

In some cases, Dance opined, Ellington's feel for "sort of eccentric voices" resulted in certain singers being added to the line-up, for instance, Hibbler, whom Ellington described as "a master of vocal pantomime." In other cases, Dance went on to say, "Ellington liked to see what he could do with a voice of very small, effective range," citing Milt Grayson's deep baritone, or voices of "particular ability," like that of the opera-trained Davis. "I guess what he really meant," Dance conjectured in an attempt to fathom Ellington's self-effacing remark, "was that he could never tell what the public would like."

Two Ellington band singers not mentioned in the foregoing but worthy of our attention are June Norton and Jimmy McPhail. That both called Washington, D.C., home has been advantageous to me, for I have had a number of occasions to see them perform, sometimes in the context of tributes to Ellington, and their proximity has lent to interviews a pleasant informality.

"Duke and I had worked the same stage before at the Howard Theatre with a lot of stars when we tried to raise money for an artist," June began in 1985. "You know how we do that kind of thing and although we're already working, we run in when we leave our club between sets. And so our paths had crossed. Jerry Rhea, who used to be Duke's road manager, kept trying to get Duke to listen to me, trying to make us know each other.

"Finally Jerry called me one evening and said, 'June, I want you to catch a cab and come over here to the Howard Theatre. Duke wants to meet you.' I said, 'Duke who?' 'Cause he was the last person I thought would want to meet me. So I went over there and the theater was closed. Duke liked to stay behind when a house is empty—that was when he created, in the wee hours of the morning.

"He said, 'Come on, let's sing a few things,' and it ended up into about an hour and a half session. Then he said, 'Well, come on, let's go to the house,' his family home here in Washington. And I guess we spent about two or three hours—Jerry and Duke and I—just singing songs.

"So Duke told me that he was going to put me under contract and was going to put Jerry Rhea as my coach and he wanted me to learn a lot of the things that Kay Davis did. Kay had recently left—this was 1951—and he wanted me to learn 'Translucency' and 'Creole Love Call' and all those things. Well, that wasn't my shot but, you know, I had had that classical training because that was where I was headed, initially, . . . to the opera. But I found out, after years of study, that I couldn't get into the Met because, you know, I'm Negro. So I pursued my voice training, trying to sound a little more popular than classical, and having had all those lessons, it took a lot of training for me to make a marriage of the two. Even after I left Howard University I studied under Frederick Wilkerson and also Dini Clark worked with me.

"But getting back to the thing with Duke, he asked me not to accept any job that would be binding to the point that it would not allow me the freedom to be on a moment's notice and so I did exactly that. I said that I was under contract to Duke and that if I were hired it would be with the understanding that, should he call, I could leave immediately. And I still studied with Jerry.

"And one night Duke called me and said, 'I want you here,' and he told me to join the band at Back Bay Station in Boston. I went there and we did all the New England states and then we dropped down into New York to do the Cavalcade of Bands that Rexall druggists used to do.

"The friendship that developed out of my having been with the band was the thing that I cherished so much. Another thing, you had to be very, very on the ball, you couldn't sit up there and not pay attention to what was happening because he never called your name, like, 'Okay, June, you're going to do this.' But Duke and I formed a close relationship because I'm a very forthright kind of person and I never tiptoed around Duke. I would speak my mind, say what I had on my mind, and sometimes it took him back because I think people kind of tiptoed around him a little bit. He raised his eyebrows at me one time and I thought, 'Oh, boy, I'm fired, I know this.' But instead, he realized that I was an honest person.

"He used to infuriate me because he knew when we'd stand in line and they would come from the tracks right on up to the bluebloods, they'd all stand there waiting to have an audience with him, and I know how he'd press that button to say almost anything and never take a breath almost. So one time he said, 'You're a beautiful person,' and I said, 'You don't have to tell me things like that because I have eyes and I can see myself very well. I am not beautiful.' And he said, 'Are you quite finished?' And I said, 'Yes,' and he said, 'I mean you're beautiful inside and that is reflected in your face and in what you do and in how you carry yourself.' So I was very embarrassed then and I said, 'Well, forgive me for jumping the gun,' and he said, 'Really, all is forgiven.' Then we became friends."

June Norton's stay with Ellington added up to about a year, and she occasionally appeared with the band after that, especially when the orchestra visited the Washington area and its leader would insist that she come up and do a few numbers. June continued to perform in clubs with a combo over the years, often accompanied by John Malachi, the former Billy Eckstine band pianist and arranger and a fellow D.C. resident.

In 1973 June earned a Master of Music Education degree at the D.C. branch of Antioch University. She went on to work with underprivileged children under the city's Department of Recreation, and she joined with former Ellington trombonist Lawrence Brown, who settled in Washington in the early 1970s, in a program that took music into local women's prisons. In the 1980s June became active in the International Duke Ellington Society, participating at its annual meetings here and abroad both as performer and as panel member.

"Duke and I had become very close," June once told me, "and I was at his death bed just before he died. Not many people were allowed in but I went up to see him and we talked. We always had shared things with each other that we didn't share with other people and I had one thing I had not told him. I wanted him to know that I had thought about it and how I felt about him and I wanted him to know

that before he left. So I told him that 'Most people walk on the surface of the earth but you don't and you haven't. You have pushed
your feet deep down into the earth and planted seeds so that all man-
kind could reap the harvest of your crop.' And I told him that I also
thought so many times how most people could not touch the hem of
his garment and I was so happy and grateful for friendship that he
had given me all of my life and I treasured every moment of his
friendship. Well, he fell silent and he looked at me very quietly and
introspectively and then he took my hands—and Duke always did the
continental thing, you know, kissing the backs of your hands—and he
turned my hands over to my palms and kissed both and I almost wept,
but I didn't want to cry because I didn't want to spoil my visit and so
I said, 'Well, I'm not going to keep you up and use up your energy'
and 'Thank you so much.' He said, 'I want you to come back tomorrow,' he said, 'and bring my book so that I can autograph it.' I said,
'Tomorrow? What time?' He said, 'Ten o'clock.' I said, 'In the morning?' He said, 'No, at night.' I said, 'Are you sure? Ten at night?' He
said, 'Yes.' I said, 'All right, I shall.'

"And I returned and brought my copy of his *Music Is My Mistress.*
Jimmy Lowe, his valet, was on duty because that was Duke's worst
time of the day—the nights were awful. So he autographed the book
and it took him quite a long time because he was having a hard time
of it. He wrote, 'Pour la plus belle, June Norton, Love always, Duke
Ellington,' with kisses underneath."

" 'Lord,' I said to myself, 'that *I* am going to have an opportunity
to sing with his big band,' " is how Jimmy McPhail recalled for me the
moment he was told that he had won a WWDC radio station talent
contest thirty-five years before, the prize for which was a week's book-
ing at the Howard Theatre with the Duke Ellington Orchestra. "I had
been going to the Howard Theatre for years and hearing his music
but I'd never met him."

For the then twenty-three-year-old singer the scene at the first rehearsal was enhanced by "seeing guys who were great names with
Ellington." Jimmy not only did that week with the band but traveled
with it for a year and continued to perform and record with Ellington
off and on until 1974 when the great bandleader died.

The Rocky Mount, North Carolina–born McPhail, who came to
Washington, D.C., at the age of three months with his parents, began
singing when he was eight in the choir of Galbraith AME Zion Church
under the direction of Alma Harris. "I just tried to familiarize myself
with what I heard," declared Jimmy, who has never learned to read
music. The singing continued through Slater Elementary School, where
he remembers doing solos at assemblies, and at eighteen he was put
into the senior choir at church.

There was no record player in the McPhail home, but Jimmy absorbed via radio the popular music of the day along with Ellington, Count Basie, Frank Sinatra, and combo jazz. "We started a group called the Armstrong Four in high school, imitating the Mills Brothers and the Ink Spots and groups of that kind and got quite popular around the city, working in little nightclubs. For about five hours a night we got a dollar and a quarter apiece."

Jimmy returned to his native state to continue his education at Shaw University in Raleigh, where he had his first big band experience.

"A classmate was working in a nightclub and told [white band-leader] Woody Hayes about me and I started singing with the band. I stayed with him for the four years I was there and I also sang with a black band in little towns around Raleigh. With Hayes's band I had to enter through the back doors of the clubs. There was no problem being on the bandstand. I would do my stint on the stage and then go back into the kitchen. I worked at the Sir Walter Raleigh Hotel in Raleigh and the same situation existed. You did all those things because that's the way it was."

After his year with Ellington in the early 1950s Jimmy worked with Ella Fitzgerald and then went out on his own, picking up gigs up and down the East Coast as a single with local bands. "But I got homesick and came back to Washington, D.C.," Jimmy confided. In 1959 he opened Jimmy McPhail's Gold Room, which in the late '80s could make some claim to being the oldest continuously operating jazz venue in the Washington area. During the first year of the club's existence, Jimmy recalls with a chuckle, "they were tearing up Bladensburg Road and it would rain almost every week. I put boards across the street so the people would not step in the mud." Redd Foxx, Etta Jones, and Al Hibbler were a few of the acts that would fill the popular nightspot during the '60s, and Jimmy was still performing there in the 1980s when he was not in New York or elsewhere lending his tenor voice to Ellington tributes.

"I'll never forget one time in Phoenix with him," Jimmy reminisced, "and we worked the Star Theatre, like a theater-in-the-round. I did a medley and the people applauded for the longest time and I stood out there waiting for Duke to tell me what to do or whether I was supposed to sing anymore or not. Shucks, he just kept running up and down the piano and I eventually backed off the stage. Talk about intimidating!"

By many accounts, including several alluded to here, Duke could be intimidating. That he also was gifted with a rich sense of humor has been often attested to. Sometimes it was expressed with dry, understated wit. The vocalist Al Hibbler, who was born without sight, first met Ellington in 1935 when he was studying voice at the Conservatory for the Blind in his native Little Rock, Arkansas.

"I had an audition with him," Hibbler told me, "and I was drinking

pretty heavy. Ivie Anderson wanted to take me and look after me, but
Duke said, "I can handle a blind man, but a blind drunk, I can't make
it.'

It was nearly a decade later that trumpeter, violinist, and singer Ray
Nance surreptitiously introduced Hibbler onto the bandstand one night
and the bandleader, upon hearing the distinctive and quite unique
vocal abilities of Hibbler, hired him. "I was in the band for two weeks
before I even knew I was in it." That was in 1943, and Hibbler remained with Ellington until 1951.

Alto saxophonist and arranger Rick Henderson, who was a member
of the Ellington reed section in the 1950s and on and off until the
leader's death, related with amusement his initiation into the great
organization. "The way Duke auditioned, he just put the uniform coat
on you, set you down in the band, and he'd go on and play as if you'd
been there all the time."

Rick also recalled, chuckling, that "One night before a recording
session the next day Duke casually dropped by my hotel room at 2
a.m. with some music for me to write by ten o'clock in the morning. I
was half asleep and I promised I would do it, but when I realized
what I was talking about, Duke going out and having a ball and I was
getting up at two o'clock and write all night long and then go on and
record, I got up out of bed and went and found Duke and gave him
the music back. And he laughed 'cause he realized I realized what he
was doing—he was planning on having a ball. So I went back and got
me some sleep."

Having given Goodman and Ellington a mere modicum of their
due through the reminiscences of several who were close to these two
great artists, we must turn our attention to another whose contribution to big band jazz was, while of a different nature, of equal impact
on the idiom. We refer, of course, to Count Basie, whose leadership
of his own big band (scaled down to an octet for several years in the
1950s) stretched across half a century, a tenure matched in the annals
of big band jazz only by Ellington.

Again we turn to a singer for background, Helen Humes, who was
with the Count Basie band from 1938 until 1942. Her story before,
during, and after this four-year period has much to tell us about our
culture.

"We didn't have a record player or anything so I didn't get to listen
to the different people," Helen explained. "So I can't say that I had
an influence. I just influenced myself when I wanted to sing. I used
to just play the piano and sing things the way I felt like singing. I
lived next door to the church and I'd play for the Sunday school, and
we had a little choir.

"Miss Bessie Allen had the Sunday school there in Louisville, Kentucky, and if you would go to her in the evenings, she would teach
you different instruments. That's where Jonah Jones learned to play

the trumpet. Well, I used to play piano with a little group with Jonah and trombonist Dicky Wells, a little dance band, and we used to play at Sunday school and at the movin' picture theater on the weekend.

"There was a man, Sylvester Weaver, he came to the theater one night and he heard me playin', and I sang there, too. Well, he wrote to a Mr. Rockwell and Mr. Rockwell came to Louisville and Mr. Sylvester brought him to hear me. So I played and sang for Mr. Rockwell and he asked my mother if I could come to St. Louis and record. So Mama said it was all right, but she'd have to bring me." (This was 1927 and Helen was thirteen years old.)

"So I went to St. Louis and I made this record, 'Do What You Did Last Night' and 'If Papa Has Outside Lovin' ' and 'The Racetrack Blues.' The kids back home would say, 'Helen, you don't act a bit enthused about it.'

"Well, in 1937 I was working in Cincinnati at a place called the Cotton Club and Basie and his band came to Cincinnati and they came out there where I was singing. Basie asked me would I like to join his band. Billie Holiday was leaving and he wanted to know if I wanted to sing with his band. I asked him how much did he pay and he said thirty-five dollars a week. I said, 'Oh, I make that here and I don't have to travel all around.'

"So I didn't pay it any attention then and I stayed there. I was working with Al Sears' band. We came out of the Cotton Club and we traveled around to Philly and Pittsburgh and what-have-you and we finally landed in New York. We were playing the Renaissance Ballroom and they were having a big affair there one night and John Hammond was there. John liked me and wanted to know if I would come and talk with Basie.

"In the meantime, they used to have Amateur Hour at the Apollo Theatre, and everybody was saying, 'Helen, go on and get it.' So I went down there one night and Basie was playin' there that night. Well, anyway, they had this contest and I won second prize. Some girl that was just note-for-note like Ella Fitzgerald won first prize. Well, I guess Basie tried her and when they gave her the music and they didn't have no Ella records around, I guess she couldn't make it. Anyway, they sent for me and I joined the band when they went into the Famous Door on 52nd Street in 1938.

"All those one-nighters was new to me," observed Helen, thinking back on the constant touring. "Yes, indeedee, one-night stands everywhere. It was fun and I enjoyed it all, but I wouldn't like to make one-nighters any more."

Helen Humes continued to work with various groups here and abroad until she returned to Louisville in 1967 to care for her parents, who were ill. She gave up singing until Stanley Dance persuaded her to come and perform in a Basie band reunion concert in Carnegie Hall in 1973.

"Well, it was during the Newport Festival and it was very successful," Helen recalled, "and about five days later they had me come over to France to work with Jay McShann, Milt Buckner, Arnett Cobb, just a whole bunch of fellows that I hadn't seen in so many years, and that was quite a ball. And what was the nicest thing was that I made an album over there and it won an award there. Then I came back and in 1974 the people from France sent for me to come back—they wanted me to make my own tour. So I went and stayed over there almost six weeks that time. So I was there in '73, '74, and '75.

"You know, when I'm in New York down at the Cookery, well, I'm just a few blocks from New York University and so many of the kids come and they just sit there and they make me feel good," said Helen, laughing. "They say, 'Oh, Miss Humes, we certainly are learning something, because we never heard anybody sing like you before.' You know, after a while I guess they get kind of tired of the loudness of rock and the same thing over and over.

"I just considered myself a singer, but so many people, they want me to be a blues singer. I sing two or three little blues numbers, but I don't like this down-home moanin' and sad blues. I usually like the little fun numbers." She could well have added that ballads were especially dear to her.

I last saw Helen Humes, only a few months before her death in 1981, performing at Charlie's Georgetown in Washington, D.C. She was in good voice and in good spirits, although she apparently was not well. But Helen was a trooper and the show had to go on.

"Basie *was* cool and laid back," was among the many insights into the character, personality, and style of Count Basie provided me by Thad Jones. The conversation took place in early 1985 about two weeks after Thad had taken over leadership of the still-touring Count Basie Orchestra (Basie had died in April 1984), a position which he held until six months before his death in August 1986. Thad also filled in his early background for me, in the process supplying a window onto the Detroit jazz scene of the 1940s.

"I think Basie's approach to life was not really like a frantic hyperenergetic type of thing," continued the multi-brass playing composer and arranger who had spent the years 1954 to 1963 with the band. "It was a phenomenal thing that he had almost total recall and he was aware of everything that was going on all the time. Not only was he aware but he was concerned with what was going on and he was a very understanding man. It was truly startling how active his mind was. He was not grabbing at you with, 'This is *my* band' and that sort of attitude. Actually, all of us, we thought of it as *our* band. That's the kind of feeling that he generated in everybody. The trumpet section, for example, Snooky Young, Joe Newman, Sonny Cohn, and I, we were like brothers. Each section had a sort of a tight bond that they self-taught one another.

"It was quite intense," was Thad's way of describing the camarade-
rie of the Basie band, "and I think that was because we not only felt
the music together but we felt ourselves as a family and that sort of
feeling was transmitted through the music. So there was a roundness
and a warmness and a togetherness about everything that we did that
was very exceptional. It was different in the sense that it was coming
from that strong and binding family circle. It was incredible that a
man could organize people to form this strong bond of friendship
and generate such a warm, human feeling toward one another, concern for each other's welfare, and consistently maintain it, as Mr. Basie
did. That's true genius. Actually, it was the most unusual circumstance that I've ever been around in my life."

Of the many cities whose histories vis-à-vis the evolution of jazz have
not been adequately documented, Detroit (and its environs) ranks high
on the list of those that deserve a volume, perhaps several. The roster
of jazz artists who came up in Detroit in the 1940s and '50s is long
and impressive, including, along with Thad Jones and his brothers
the pianist Hank and drummer Elvin, trumpeters Howard McGhee
and Donald Byrd, trombonists Frank Rosolino and Curtis Fuller, reed
players Rudy Rutherford, Yusef Lateef, Lucky Thompson, Billy
Mitchell, J. R. Monterose, and Jack Montrose, pianists Tommy Flanagan, Barry Harris, Roland Hanna, Kirk Lightsey, and Bob Zurke,
guitarists Everett Barksdale and Kenny Burrell, bassists Paul Chambers, Ron Carter, and Major Holley, drummers Louis Hayes and Oliver
Jackson, vibraphonists Milt Jackson, Dave Pike, and Terry Pollard (who
also played piano), harpist Dorothy Ashby, and vocalists Betty Carter,
Sheila Jordan, Slim Gaillard, Herb Jeffries, and Barbara Dane. We
devote space to such a catalogue because it is an impressive one and
because, while several of those cited turn up in other contexts in this
book, we have not included a chapter on the Detroit jazz scene.

Returning to Thad Jones, let us allow him to flesh out that scene.
"My whole childhood was spent right there in Pontiac, Michigan, and
my parents died right there. In fact, the family residence was main-
tained right up until the time of my younger sister's death earlier this
year. I have two sisters still alive and at one time they were both musically inclined. My older sister, in fact, was teaching piano up until
the time she had a heart attack several years ago.

"There was always the sound of the piano in the home and, naturally, the sound of the radio, which was probably our main source of
entertainment back in those days," said Thad, who was born in 1923.
"I was made acquainted with the music of Ellington and Lunceford,
the early Count Basie, Woody Herman, Jan Savitt, Washboard Sam,
hosts of blues singers, and all sorts of programs that really stimulated
the imagination to a great extent, you know, that gave you the feeling,
just from listening to them, that you were right there with them. There
was a lot of symphony music, especially on Sundays. Sunday seemed

to be a day that was devoted mainly to symphonies, specifically the Detroit Symphony on Sunday evenings. During my school tenure I was exposed to quite a number of them—Mozart, Haydn, Beethoven, Tchaikovsky, people like that.

"I started playing when I was thirteen and Armstrong was my major inspiration. I thought there was nothing else more exciting than the sound of Armstrong playing the trumpet. Naturally, I wasn't too aware of the kind of technical facility that he had acquired to enable him to do all those things that he could do, but later on, as that actually began to sink in and after I began to try to learn the trumpet myself, it dawned on me that this is kind of hard.

"I got my first trumpet from my uncle, who was a trumpet player. I can still see it. It was an old Conn horn that didn't even have a case with it. I used to have to take it to school in a paper sack," Thad laughed, "and there were several embarrassing moments, especially during the winter and I had to sit there and wait until my valves thawed out before I could get up on the bandstand with the rest of the guys.

"The closest we came to a family band was a band that we organized in our hometown of Pontiac, which was actually organized by my uncle. My uncle was one of the trumpet players and I was another. Hank was the piano player. That particular band had quite a few other people involved in it and everybody, strangely enough, with the exception of the drummer, was in some way related.

"After I'd gotten back from the army in 1946 I began to migrate in the direction of Detroit, which was my natural inclination because there was not too much action going on in Pontiac. Around 1949 I started to work with different people. I started off with a guy named Candy Johnson and I did a short stretch with Yusef Lateef. Lateef and I eventually wound up at the Bluebird, which was a very popular jazz club on the West Side of Detroit. Billy Mitchell had the band over there, five pieces. Elvin and I worked in that small group together with Billy. There were three piano players that all worked there with Billy at one time or another—Terry Pollard, Barry Harris, and Tommy Flanagan.

"Jack Teagarden's son, Jack Teagarden, Jr., had a band that I was with in the army, and I was in another army band that was traveling with the G.I. show called Bedtime Stories and I did some arranging with that band. And we had some experimental bands around Detroit. But, consistently, I hadn't done anything with big bands up to the time that I joined Count Basie in 1954. That opened my ears to what big-band playing was all about."

Thad contributed many solos to recordings by the Basie band during the nine years he spent in the organization, but one cut stands out, the title track of the 1956 album *April in Paris*. The number became a Top 40 hit as a 45 rpm and the solo was later orchestrated for the entire trumpet section.

"We had been involved in the recording of this composition for maybe twenty, twenty-one, twenty-two takes, during which time I was thinking to myself, 'Gee if we're going to make a record, I better try to put myself together and say something halfway decent,' " Thad reconstructed for me the session in the studio three decades before.

"Each time we did a take, I would come up with a different solo that Norman Granz, our producer, didn't seem to care for. There was always something just a little less than what was required, in his mind. At any rate, by the time we got to, I would say, about take twenty-two, I was a little tired and I told Snooky, I said, 'this time I'm just going to go ahead and play the first thing that pops into my mind.'

"So when we started the take and the time for my solo came around, I just started playing this 'Pop Goes the Weasel' bit and when we finished, Norman said, 'That's fantastic! That's the take, we'll take it!' So I guess it had some appeal."

Thad also expatiated upon some of the guidelines that had shaped his approach to his art and he spoke of the high hopes that he had for the future for the Count Basie Orchestra, which he anticipated being at the helm of for some years.

"If you don't know the melody, it's very difficult to relate to the tune that you're playing. So the melody is primarily important, it's the thing you learn first. Then you learn the harmonic progressions that go along with it and you can develop that as far as you like and go into any rhythmic phase that you like to. Ben Webster was very fond of playing melody and I don't think there's any tenor player alive today that could play a melody greater than Ben Webster. And one of the things that he felt that helped him to really relate to and understand the piece he was playing was by learning the words to it. Know the melody and the words. Without those two things, I don't think you can really find anything of substance in, especially, ballads.

"The band is pretty much intact from a year ago when Basie died and, as far as I can see, the band's spirit is beautiful. It's warm, it's strong, it's forceful, and it's a spirit that's on the move and it's going forward. And with all those soloists they have up there now, it's rather intimidating."

The Count Basie Orchestra visited my turf in March 1985, and my then eleven-year-old son Sutton and I checked it out. Here is my review of the evening for the *Washington Post.*

The Count Basie Orchestra was without its late, legendary leader last night at The Barns, Wolf Trap Farm Park, but it was not without the high spirits, casual yet precise section work, and exhilarating swing that marked it during its nearly half-century un-

der its founder. Charts by Sammy Nestico, Ernie Wilkins, Frank
Foster, and Quincy Jones, topnotch soloists like tenor saxophonist Eric Dixon and Kenny Hing, altoist Danny Turner, trumpeter Sonny Cohn, trombonist Bill Hughes and flugelhorn player
Johnny Coles, along with the dynamic conducting of cornetist
Thad Jones, make the twenty-one-member band the number one
unit of its kind playing today.

No one could fill Basie's shoes, but pianist Tee Carson's subtle
evocation of the late leader's laconic keyboard approach saun-
tered with style and dignity, successfully avoiding caricature and
achieving inspired re-creation. The 4/4 of Freddie Green's guitar, the steady bass of John Williams, and the powerful beat of
Duffy Jackson's drums kept the band "swinging you into bad
health." An earthquake of a solo by Jackson and vocalist Carmen
Bradford's down home prescription for the blues on "Dr. Feel-
good" were bonuses in an evening already overflowing with re-
wards.

"Total commitment—I think that, too, is very important," was how
Thad had concluded our interview of a few weeks before this concert.
"I believe that you should play as though there's no tomorrow, that
you should devote yourself totally to whatever you're doing, because
you may never see another day or another moment." Thad Jones died
in Copenhagen, where he had directed the Danish Radio Orchestra
during the '70s. He had completed his one-year contract with the Basie
band in February and had continued to play until cancer was diag-
nosed four months before his death.

Having heard so far in this chapter from several artists whose associations with Basie, Ellington, and Goodman cast light upon the modus operandi of each of those great bandleaders, let us hear directly
from a player of the same generation, one whose career was still in
progress in 1990, Lionel Hampton. Hamp's life story, related to me
by the vibraphonist, drummer, bandleader, and composer, is a remarkable one, not just in terms of the music but as a black counter-
part to the Horatio Alger theme in American socio-cultural history.

For more than half a century, the figure of a sweat-dripping Lionel
Hampton leaning over his vibes or in furious motion at his drum kit
has loomed large by virtue of his many film appearances with the Les
Hite band (in the 1930s), subsequent cameo roles in movies that featured numbers of his own and others' bands (some of which films
turn up on television with regularity), and his more recent television
activity. Hampton is, in fact, one of the several best-known visual
presences of the jazz world. Whether mercilessly whipping a jitterbug-
packed dance floor into action in the '40s or leading a supercharged
young band at an SRO club engagement in the '80s, Hamp's pre-

eminent skills, charismatic showmanship, and sheer energy have rarely been matched in the performing arts.

In the early '80s no less a percussion master than Louie Bellson engaged in a drum battle with a septuagenarian Hampton at Wolf Trap Farm Park in Vienna, Virginia, sat back, and, beaming admiration, conceded defeat. At a 1983 concert at Washington, D.C.'s Howard University, where he holds the rank of visiting professor of music, Hampton was the central character in a star-studded cast that included Dexter Gordon, Teddy Wilson, Milt Hinton, and others with whom he has been associated over the years. The event was a retrospective in miniature of Hampton's career. There were explosive re-creations of the Benny Goodman Quartet's "Limehouse Blues" and "Nagasaki" along with big band settings, with Gordon's big-toned, visceral tenor supported by the leader's driving vibes.

Throughout the concert Hampton variously scatted, mugged, and improvised shuffling dance steps to the delight of the audience, mostly made up of students, few of whom had ever witnessed him in action. The sight of a man in his mid-seventies cutting up like a twenty-year-old was no doubt a revelation to many. But for older hands, it was the familiar Hamp, whose artistry and antics never fail to entertain because they spring from a man who works hard as he has fun.

In the interview two years later, Hampton offered his own assessment of the factors—especially his determination and ability to seize every opportunity—that carried to fame and fortune a poor youngster from Louisville, Kentucky.

"Well, I tell you," he explained, "you had to find the way, you had to have a good initiative. If you're looking ahead, and you want to go ahead, you know, if this gave you a little inch, well, you take a foot."

Lionel Hampton, born April 1909, spent his youth living with his mother and grandparents in the black sections of Birmingham, Alabama, and Chicago. His father was long believed to have been killed in World War I, but Hampton told me that years later he tracked him down in a veterans' home in Dayton, Ohio, where he died in the early '40s. "He was supposed to be a piano player and my mother sang in church," Hamp recalled.

His parents' musical bent and his own exposure to the music of the black church were building blocks in Hampton's early musical education. But it was at the Holy Rosary Academy in Kenosha, Wisconsin (where Hampton was sent because of his grandmother's concern over the school conditions in Chicago), that the future drummer received his first real musical schooling.

"There were about five or six hundred boys there and a Franciscan father, the Rev. Steven Eckhardt, who had a mission to help black boys and Indian boys," Hampton pointed out, "and Sister Petra was the mission's Dominican sister. She started a drum, bugle, and fife corps at the school and I applied for drums. Before then I was beating on my mother's pots and pans and using two rungs off a chair for drumsticks. So now I got a chance to play drums and this sister was our teacher. Boy, was she great, too. She was hard on you—if you didn't play those paradiddles right, she'd take those sticks and break your knuckles. We had to practice day and night. But when I came out of that school I really knew those paradiddles, like I know 'em today."

If the spare-the-stick-and-spoil-the-drummer pedagogy of Sister Petra instilled the rudiments in young Hampton, it was a slightly later experience that made him the musical sophisticate who would one day compose a suite for a guest performance with Arthur Fiedler and the Boston Pops. Returning to Chicago to continue his education at St. Monica's School, located on the South Side, Hampton "used to hear music played out of this building at 37th and Michigan Avenue." It turned out to be the *Chicago Defender*'s Newsboys Band that had caught his ear. In a matter of days he met the entrance requirements of the organization by landing a position "peddling papers." The band's director was Major N. Clark Smith, a distinguished black educator whose students over the years included, besides Hampton, bandleaders Bennie Moten and Harlan Leonard, bassists Milt Hinton and Walter Page, violinist Ray Nance, and many other jazz greats.

"Major Smith liked me, so'd I'd play drums in the marching band, and in the concert band I played timpani, xylophone, and orchestral bells. Major Smith had studied in Heidelberg Germany and was a specialist in solfeggio harmony and he taught me that." Hampton credits Smith with providing the ear training that later, enabled him to transfer the solos of Louis Armstrong, Benny Goodman, and saxophonist Coleman Hawkins to his set of orchestral bells, then to the xylophone and the vibraphone.

Hampton's routine while a busy member of the ninety-strong Newsboys Band included daily band practice, playing in parades and at special parties, and, of course, peddling those papers. Somehow he found time to check out bands at the Vendome Theatre. His instant favorite was Erskine Tate's Orchestra, which included drummer Jimmy Bertrand, whose use of the xylophone first inspired Hampton to take up mallet instruments.

"I wanted to play anything that had a keyboard like that," confessed Hampton, and the persistent youth soon persuaded the older drummer to give him some lessons. Bertrand, whose other students included Big Sid Catlett, was adept at tossing his sticks in the air and catching them without missing a beat, and his drums sported flashing lights, gimmicks that led Hampton years later to don fluorescent gloves and, with lights out, do his "ghost drummer" act.

At the age of fourteen Hampton received a set of drums—flashing lights and all—from his grandparents, and was soon playing with local groups. A fellow member of one of these units, saxophonist Les Hite, went out to California and in 1927 sent for Hampton. "It happened my aunt was going out the same time, to be some kind of helper for a rich white woman, I think, so I begged my grandmother to let me go with her. She said she would, if I would finish school."

The eighteen-year-old Hampton who arrived a few months later in Los Angeles certainly had little idea of what was in store for him. He served in the bands of Hite, the Spikes Brothers, Vernon Elkins, and Paul Howard, with whom he made his recording debut as a drummer in 1930. That same year, as a member of the house band at the Cotton Club in Culver City, California, Hampton took a giant career step that would establish him as a musical pioneer. Louis Armstrong arrived for an engagement at the club without an orchestra and was so taken by the house band's backing that he took them into a nearby studio to record.

"We got to the recording session and there's a set of vibes sitting in the corner," Hampton related. "They weren't playing nothing more than the 'N-B-C' at that time on the vibes, just making whole notes. So Louis says, 'What's that in the corner?' I said, 'Oh that's a new instrument and it's part of a drummer's outfit and it's called the vibraphone.' So Louis said, 'Well, play something on it.' So I moved 'em out on the floor and played one of his solos on the vibes and Louis had a fit about that. He says, 'Oh, man, let's put it on a record.' Eubie Blake had just sent Louis an arrangement of his "Memories of You" and I played the introduction and played all through it. So that was the first time that jazz had been played on vibes on a record." Hamp soon had his own set of vibes, which gradually supplanted the drums as his main instrument.

In 1934 Hampton formed his first band, which for a time included Buck Clayton and tenor saxophonist Herschel Evans, both of whom would later be stars of the count Basie band. After playing clubs along the West Coast, Hampton settled at the Paradise nightclub in Los Angeles. Two years later fate struck again. As he recalled, "It was a place where all the sailors came to drink the twenty-five-cent beer. I had a ten-piece band with Tyree Glenn and Teddy Buckner—the best jazz band in town. Lo and behold, I was playing one night and I heard this guy playing clarinet beside me and it was Benny Goodman. It happened that Benny was playing the Palomar Ballroom and John Hammond is supposed to have told him about me. That's when I first ran into Benny Goodman."

The next evening Goodman brought Teddy Wilson and Gene Krupa by to jam. The upshot of these impromptu sessions was an invitation to record with the Goodman Trio. They cut "Moonglow" on August 21, 1936, and several days later returned to the same Hollywood studio to record three more tunes, "Vibraphone Blues," "Exactly Like You," and "Dinah." Soon after this date, before the records came out, Goodman called Hampton from New York. The Goodman Trio was about to become a quartet.

"I didn't know if I was talking to Benny Goodman or not," said Hampton of that fateful moment, "but finally Tyree Glenn called my wife Gladys to the phone and she talked to Benny and he wanted us to fly to New York. But Gladys said, 'No, we'll drive.' She had a little white Chevrolet and she got a trailer and we put my vibes and drums in it and we got to New York on the 9th or 10th of November.

"I made my first appearance with Benny Goodman on the 11th in the Madhattan Room of the Pennsylvania Hotel. It was instant integration, the first time that black and white ever played together on stage, 'cause Teddy was traveling with Benny, but he wasn't playing with him on the stage, he only played in the intermissions by himself, see. The group wasn't integrated until Benny got the quartet. Black and white wasn't playing together in baseball, basketball, football, and blacks only had a little part in motion pictures and had to use that funny language, 'Yes'm' and 'No suh.' "

When Goodman gave up his band temporarily in 1940 because of ill health, Hampton seized the opportunity to organize his own big band. "We had some fantastic players," he boasted, referring to a roster that by any standards constitutes a veritable *Who's Who* of jazz greats. Among those musicians were vocalists Dinah Washington and Joe Williams, trumpeters Clifford Brown, Fats Navarro, Art Farmer, Kenny Dorham, and Snooky Young, trombonist Kai Winding, saxophonists Johnny Griffin, Dexter Gordon, Charlie Parker (on tenor), and John Coltrane (on alto), guitarist Wes Montgomery, bassist Charles Mingus, pianist-arranger Milt Buckner, and trumpeter-arranger Quincy Jones.

So astonishing are Hampton's achievements over the course of his more than six decades of professional activity that one wonders how a single lifetime could contain so much. Not only was he a trailblazer on the vibraphone and a master showman, he was also the first bandleader in jazz to use the electronic organ and electric bass guitar on a wide scale. He also produced nearly one hundred small group recordings in the late 1930s and early '40s, which in critic Stanley Dance's words "provide a remarkably panoramic view of the state of jazz during the period of its greatest popularity." And he was one of the first black public figures to buck the tide of racism in the mid-1930s.

Hampton's talents have been duly recognized with honorary university degrees, a Papal Medal from Pope Paul VI, and the New York Governor's Award. But one honor in particular stands as a milestone in his career: a certificate presented to him by Broadcast Music Incorporated (BMI) at the 1978 Newport Jazz Festival, marking the one millionth performance of "Flying Home" of which he was co-composer (with Goodman) and which has long been his theme song. The myriad renditions of a tune so long associated with him constitute a symbol of the reverence for Hampton held by the public and fellow artists alike. It's difficult to listen to "Flying Home" without imagining Hamp's mallets racing across the metal bars of his vibes. In fact, seated in the concert hall at the 1988 North Sea Jazz Festival in The Hague, Holland, I didn't have to imagine—there was Hamp fronting a big band of musicians half and even a third his age. And he was pushing those young cats to their limits!

Another star of the Big Band Era and one whose duration as leader has been surpassed only by Ellington, Basie, Woody Herman, and Hampton, Harry James celebrated his forty-fourth year at the helm of his own band in January 1983, half a year before his death. I interviewed the great trumpet player in 1982. At that time he was still spending eleven months of the year on the road, traveling the length and breadth of this continent, and fitting in tours to Europe, South America, and Japan. He pointed out to me that that was all he had ever known.

"My dad was a circus band director and my mother was the prima donna and aerial performer," Harry told me. "I was born in Albany, Georgia, and my mother and I left there when I was thirteen days old and rejoined the circus. We traveled all the time, went all over the U.S., Canada, and Mexico, nine months out of the year, and then we would have the three months off, what they called 'winter quarters.' I didn't know anything else.

"My dad started to teach me trumpet when I was nine years old and by the time I was eleven I was playing in the circus band with him. My idol at all times and my guide was Louis Armstrong. As far as I was concerned, there wasn't anyone but him. Of course, in those days you only had two choices, you either had to like Armstrong or Bix Beiderbecke and I went into the jazz school and the blues school from Armstrong. Of course New Orleans was only about three hundred miles from where I was brought up in Beaumont, Texas.

"I was groomed to be a trumpet player, a legitimate trumpet player, and I won the state contest, the Victor Herbert Contest, in Texas. I didn't start playing in dance bands until I was around fourteen when I started playing in local bands." Among the bands that Harry played with in those early years were those of the legendary Texas pianist Peck Kelly and Herman Waldman, who hired the young trumpet player for his first job with a traveling band. In 1935 Harry joined the band of drummer Ben Pollack and stayed for a year before joining Benny Goodman in 1937. January 1939 saw him on the road with his own band, with which he remained active until a month before his death.

"We take three trips a year, sometimes four," said Harry, outlining for me his band's typical 1980s schedule. "Like this particular U.S. tour is for forty days and then for six weeks we go overseas, usually in September or October—South America or Europe—and then we take a southern tour every February or March. We go down to Florida. Then in April and May we usually take another tour which takes us to the East Coast. We were in Japan twice."

Of his band's repertoire he explained, "We do practically everything that we've recorded that's been popular. We have medleys of a lot of the hits and we do a lot of today's things. We still play things like 'Two O'Clock Jump' and 'You Made Me Love You' and tunes like that which the people want to hear and we do some of the newer things like 'Pink Panther Theme' and 'Sanford and Son Theme' and things by Quincy Jones, things like that. If I like a tune, we play it. If I don't like it, I don't play it."

The trumpet and cornet player Pee Wee Erwin was a member of several big bands, including those of Joe Haymes, Isham Jones, Ray Noble, and Benny Goodman, before joining Tommy Dorsey in 1937, and later led his own big band before settling into combo work from the late 1940s until his death in 1981. I taped an interview with him for National Public Radio at the Manassas Jazz Festival, and in the course of our conversation the names of Dorsey, Goodman, and several other luminaries of the Big Band Era came up.

"I was with Tommy about two and a half years and enjoyed every minute of it for quite a number of reasons. Tommy was a very interesting man to work with, musically. He was always seeking, always looking for new things. His band was changing at all times. He was always seeking fresh ideas," insisted Pee Wee. "Sometimes he'd hire new people to orchestrate for him or to play with the band, and as a consequence it was very interesting for all of us. I think he was the first man *I* worked with that looked for, I won't say obscure people, but the lesser known people such as Willie The Lion Smith, who I don't think had a great national reputation. Of course, he did among collectors. Tommy orchestrated some of Willie The Lion's music and I don't know whether it's widely known, but Tommy was always looking for new orchestrations, although he had marvelous orchestrators.

"He had Paul Weston and Alex Stordahl as principal arrangers when I was with the band, yet he went out and looked for Deane Kincaide. In fact, he heard of a younger arranger from New Jersey—that was about the time that Glenn Miller was starting *his* band—and Tommy asked this young arranger to bring him an arrangement, which he did on 'Lonesome Road' and Tommy recorded that number. I recorded it with him. The arranger's name was Billy Finegan. And of course Tommy was well stocked with orchestrators and he liked Billy Finegan very much so he called Glenn Miller and said, 'Here's a good orchestrator for you.' And I'm quite certain that Billy Finegan made quite a number of orchestrations for Glenn Miller. Just to show the insight of Tommy Dorsey and his good judgment—it's history now— I'm talking about Billy Finegan of the Sauter-Finegan Orchestra. So of course he was an extremely efficient orchestrator. And of course everyone knows that Tommy—I don't think *stole* him away—acquired the services of Sy Oliver from Jimmie Lunceford and that's musical history."

I could not resist inquiring what was the source of Tommy and brother Jimmy Dorsey's notorious public quarrels, often on the band-stand and sometimes erupting into fist-fights.

"I think that must have started in infancy," Pee Wee conjectured, laughing. "And they had a younger sister and she was quite a good cornetist. I wouldn't doubt that there was probably rivalry between the two brothers and the sister, knowing the family."

I asked Pee Wee what kind of a person Benny Goodman was to work for and if the great clarinetist had been a perfectionist back in the 1930s when Pee Wee had been with him.

"His first major engagement—I guess it could be classed as that—was a program for the National Biscuit Company on NBC in 1935 for the "Let's Dance" program. I had the pleasure of being a member of his band at that time. Of course I enjoyed immensely working with Benny's band on those early engagements. Goodman had quite a reputation at that time as a performer. He'd made beautiful recordings on his own in Chicago in the late 1920s. The first recording that I was aware of, that I heard in the very beginning, was a 1928 trio recording he made with Mel Stitzel, piano, and Bob Conselman, drums, of the old classic 'That's A Plenty,' which is still to this day, as far as I'm concerned, a masterpiece.

"In the days that I spent a lot of time with Benny in his early days as a bandleader, I think he was largely interested in his own performance and—you mentioned the word 'perfectionist'—I think every instrumentalist is in essence a perfectionist. We *must* be perfectionists, or at least make the attempt, else—. You never stand still; you either progress or you go backward. Almost everyone strives for perfection, I'm sure."

My observation that he replaced Bunny Berigan in the Goodman band in early 1936 elicited the rejoinder, "That's a big order, to say that anyone that I've ever known could replace Bunny Berigan. I don't think anyone ever could. Everyone knows he was one of the greatest trumpet players who ever lived. I don't think to this day I've ever heard a sheer sound on the instrument which is as impressive as Bunny Berigan's. And I don't really and truly believe it was ever captured on wax properly. It was just an overwhelming sound on the instrument, to say nothing of the wonderful ideas that he played."

And what sort of a guy was Bunny personally? "Oh, I think about like the rest of us. People are just people or just musicians or whatever they are, if they're different than other people. Of course, you can have admiration for a person and they're great in your eyes whether he be a musician or—it doesn't matter what his field happens to be, he's a nice person."

That final phrase, by the way, is one that I would without hesitation use to describe the essence of Pee Wee Erwin the man, apart from his artistic persona, which I consider to be of a very high order. I can recall almost as though it had been only a year or two ago my first meeting with Pee Wee forty years ago when he had just begun a several-years residency with his newly formed combo at Nick's in New York. Brash college freshman that I was, I asked him to play, in honor of my home state, "Maryland My Maryland." Pee Wee complied with my request. The next afternoon, walking down Sixth Avenue that bright June day, I heard whistling behind me. It was Pee Wee, rendering a solo version of "Maryland My Maryland," once again for my benefit. Pee Wee, accompanied by wife and child and carrying easel and other painting paraphernalia, was enroute to Central Park to, as he explained, "sketch seascapes."

Johnny Desmond had been featured vocalist with Glenn Miller's Army Air Forces Orchestra in 1944, having earlier sung with the bands of Bob Crosby and Gene Krupa. "Glenn Miller was a very strict man, but he was always fair," Desmond assured me, a year before he died in 1985. "He knew what he wanted and he knew how to get it. He was a wonderful guy, loved playing sports. When he was out on the sports field he was one of the guys, but the minute he stepped back on the bandstand he was the consummate leader.

"We were stationed in London when we first got over there," Desmond recalled, "in Buzzbomb Alley, as they called it, a radius of three blocks of attached housing. Miller went to Eisenhower's headquarters and said, 'We can't do our work because every time one of these buzz-bombs comes the sirens go off and we have to run for the shelter.' As a result we were relocated in Bedford, seventy miles north of London. We moved up there on a Sunday afternoon and the very next morning at about six o'clock a bomb destroyed that whole block where we had been quartered and we would have had it.

"Miller was always worried about flying and he sort of had premonitions. He said to me one time, 'I don't know if I'm going to make it home.' He had just become a major, got his gold leaf. He wanted to get to Paris a day ahead of the orchestra and set up the theater because we were supposed to do a concert on that particular night. He wanted to make sure everything was in great shape when we got there. He was always that way. On details he was just fanatic.

"We had had a nice party the night before and this colonel who had his own private airplane—a warrant officer used to fly him—happened to be at the party and Miller said, 'Gee, I'd like to get to Paris tomorrow.' The colonel said, 'Well, I'm going to Paris tomorrow, why don't you fly with me?' If Miller had gone to London to requisition a flight, he would have still been alive because nobody flew for three days. That was during the period when the Germans were trying to break through up there in Belgium. Miller would never have gotten into the air because everyone was grounded for three days.

"Well, the colonel, the pilot, and Miller took off from Bedford the next day and that was the last anybody every saw them. When we got to Paris three days later, nobody knew that he was missing. It was an unscheduled flight and we were the only ones that knew about it. When we got there we sat around at Orly Field waiting for him to show up. Our captain went over to the tower and asked them where Major Glenn Miller was and was told, 'He didn't come in here. We've been shut down for three days, same as England.' " The single-engine airplane carrying trombonist and bandleader Glenn Miller is assumed to have gone down in the bad weather of that morning into the English Channel.

"He's bigger now than before he died—or disappeared," Desmond mused. "Wherever you go, all around the world, you can't escape the Miller sound. We're seeing a whole new generation of appreciation for big bands and it's wonderful to see them in front of the band-stand, just like their folks used to do. I think the reason nostalgia is so strong today is that people want to go back to a time when life was a little more simple."

Johnny Desmond came up in a time that is thought by many to have been "more simple," having grown up in Detroit in the 1920s and '30s.

"Detroit was a very exciting town in those days," he reminisced, "with a lot of places to work. A lot of well-known people came out of Detroit. It was a wonderful showplace town. We had all kinds of amateur contests where if you won you could end up with a week's work and that was the way we got our experience. I'm the only guy in my family that did anything with music. My daddy had a grocery store, a mom-and-pop store, and I used to sing around there all the time while I was setting up the fruit stands. We had this customer whose son was singing on the radio locally, and she asked my dad, 'Who was that singing?' 'Oh, that's my son John,' my dad answered. 'He likes to sing all the time.'

"Well, this lady helped me get on Uncle Nick's children's program. I had an audition and was on the air the same day. I was too excited to get scared." Desmond went on to voice and piano training at the Detroit Conservatory, which he said "served me very well all these years." He added, "I never thought about a career in those days—I just sang because I felt like it."

One of the stranger career stories of the Big Band Era is that of clarinetist Artie Shaw, whose band achieved a popularity second only to Benny Goodman's and which in jazz terms was in some respects the equal of it. Shaw organized his first big band in 1937 and was still leading a band in 1954 when he abruptly retired from music and became a writer, by his own account never playing the clarinet again. In 1983 he again organized a big band and put it under the leadership of reed player Dick Johnson. Shaw was making appearances with this band during the early months of its existence, and I took advantage of its spring 1984 one-night stop-over at the Shoreham Hotel in Washington, D.C., to interview the former bandleader. Naturally the

interview took two principal tacks: why had he so suddenly left the
music business thirty years before and why was he re-forming a band?

"You reach a certain place in your development and if an audience
won't let you express yourself as you must," Artie began, "anything
else you do is fraudulent. I just didn't want to continue standing up
there in front of a band doing things that I didn't particularly want
to keep doing.

"Out in California I've been doing a series of lectures at colleges
and junior colleges. I do one seminar called 'Three Chords for Beauty
Alone and One to Pay the Rent.' I'm perfectly willing to pay one, or
even two, for the rent, but when it gets to be three and four for the
rent, there's very little left for me. That's really the basic reason why
I quit. I didn't like the life, I didn't like the straightjacket I was placed
in. It's one thing to be a musician, another thing to become a celebrity,
and I couldn't seem to handle both of them—I don't think anyone can.

"I was color-blind and I hired people for what they did," said Artie,
in whose band at various times were such black players as Hot Lips
Page, Roy Eldridge, and Billie Holiday. "There are really two basic
perceptions you have to assume—one is of the performer and the
other is of the audience. The response of an audience is not always
fitting to what it's responding to. People's reactions are unpredictable.
I learned a long time ago to handle what is in my own hands, my own
purview. I could hire the musicians, rehearse the band, get it to sound
like what I thought it should. I could not deal with what the audience
did, and frankly I didn't care.

"If the music is awfully good and I like what I'm doing, and the
audience will allow me to do what I want to do, they don't have to
understand, all they have to do is let me do it. Then I don't care about
how much money I get. I'm a happy guy if the music works well. If
the music works badly, I don't care how much money I get. It's of no
interest to me and I don't want to be there. Now I don't want to say
that I'm totally oblivious to money, but I don't want it to dictate what
I do. It's secondary. Music is primary and money is the way you keep
score.

"So that same thing applies to what you just asked me," Artie continued. "The music was primary and what I did with the band was
what concerned me, what I did on the bandstand. . . . And that's all
I'm concerned with. I'm cursed with a sort of maverick need to chase
after an illusive thing called perfection. You're never going to get it,
but at least try to approximate it.

"That's the reason I'm doing this," said Artie, alluding to his new
musical project. "We've got a helluva bunch of musicians and if I weren't
totally satisfied that they could play the music that we used to play
plus right on up to today, I wouldn't be here at all, I wouldn't continue. I tried it as an experiment—let's see what happens. I found

that the old charts that I wrote way back then—'Rose Room,' 'What Is This Thing Called Love,' the jump things, the jazz things—they sound better than they ever did 'cause these guys play them with a kind of modern sensibility. They're contemporary musicians. My basic job is to get them to play the notes in a way which does not violate the spirit in which those notes were written. In jazz notes are only a means to an end and the aim is to get past the notes of the music.

"Jazz is an oral rather than a written tradition. You have to write the notes down so they'll know what they're playing, especially when you write for big band, but you want to get them to play the notes in a way that no one could write them. I have to sing it to them. When I used to play I would play it for them. Eventually it sinks in. The phrasing we're doing now is totally different from what we did back then. It's a lot looser and a lot more, I guess you'd say, flowing. When we were playing back then we'd play syncopation—'dot, dot, dot, dot.' Now it's 'dah dot dah dah.'

"What I'm doing is what a parent does when it teaches a child to grow up, walk, and learn something about life. You give as much as you can to it and hope that what you give will be imparted in such a way that the person grows rather than is stifled by what you do. I'm trying to teach these guys to be themselves, but through my particular view of what the music should sound like. I don't want them to be straightjacketed any more than *I* wanted to be straightjacketed. I don't want Dick Johnson to play those notes the way I did—he couldn't. Nobody can. Nobody can do as you do. You can't do what somebody else does. If you do, you become a clone. Some of the tunes we play, you'll recognize the general shape of them, but the sound of them, the way they're playing them is totally modern sensibility.

"I'll be with the band on and off. I don't want to go every night because I want them to walk on their own. When I'm there I tend to hover over them, and while I help, I also impede their own growth. My aim is to get this band to sound as if my band had stayed on into the '80s. See, they wouldn't let me do that back then. We used to play a lot of ballads and they became so famous and so successful that people overlooked the fact that I had a helluva good jazz band. I finally went out with a small group because I could play the smaller and more sophisticated audiences."

Artie was pleased with the reception the band had been receiving during the several months it had been on the road. "So far the audiences have allowed us to do what we want to do," he enthused, "and if they continue to do so, we're going to do the best we know how. Nobody has to worry about me doing less than my best."

I caught the new Artie Shaw Orchestra that evening and was listening closely for the "modern sensibility" its founding father had cited as the essence of the band's approach to the old charts. The engagement was a dance sponsored by a local big band society, and the ballroom was filled to overflowing with the organization's largely white, middle-age, and older members who, one suspected, came to hear what those old charts sounded like rendered by an orchestra comprised, for the most part, of young Berklee School of Music graduates.

Shaw conducted the opening set, offering "Rose Room," "Nightmare," "Moonglow," "Frenesi," and other hits of yesteryear in a sort of chronological order with introductory remarks on each selection. That "modern sensibility" peeked through on solos now and then, but one was left with the impression that the charts were being followed rather closely. Not that the band didn't swing, but one tended to reserve final judgment until next time around when, in Shaw's words, "they shake down and don't have to read notes any more."

The potential of the full orchestra was verified on the two tunes on which band-within-the-band units were showcased. A re-created Gramercy Five cut loose on "Summit Ridge Drive" and an eight-member combo stretched out on "Milestone." The otherwise crowded dance floor was emptied for these two selections, many gravitating to the bandstand to gawk at the swinging quintet and octet. I concluded that when the full band had thrown away its charts and begun to take it off the top of its collective heads, as these two combos had, that's when the action would start.

Speaking of that phenomenon, the band-within-a-band, which was a feature of most of the units we have included in this chapter, here are some pertinent remarks by two who participated in such combos.

Trumpeter Billy Butterfield, who was a member of the Bobcats of the Bob Crosby band in the late 1930s and Artie Shaw's Gramercy Five in the early '40s, spoke with me about the latter.

"It was a *marvelous* little band," said Butterfield, who died in 1988. "We did a lot of woodshedding. By woodshedding I mean we rehearsed a lot, we worked on it, and every number that we did was very well prepared long before we did it. It was very organized before we ever got onto the bandstand. You know, Artie was a perfectionist, like Benny Goodman. Both of them were perfectionists. Artie was a really good guy to work for. He treated you very well. He told me one time, he says, 'You know, to be a player you gotta practice and live with the horn at all times.' And he didn't want to be that strict of a player any more, and rather than tryin' to do it halfway, he just said, 'I'll do it no way,' and he became a writer.

Butterfield's early history is instructive, He was born in Middletown, Ohio, in 1917. "I started on the violin very early, five years old, and I finally went to the Conservatory of Cincinnati, studying violin. But I wasn't too good at it. I didn't really like it, it wasn't my horn, so I changed and became a trumpet player when I was about ten. I studied with Frank Simon at the conservatory. He was a great player of the Sousa school of trumpet players. I was always playin' at jazz, I just gravitated to it, it was almost natural. Louis Armstrong was the great influence on my life. There were little bands around, you know, and I was playing with a band when I was a child. My father used to be with me on the job."

If the Big Band Era per se can be defined as the decade from the mid-1930s to the mid-1940s, a period when swing was the popular music of the land, it is nevertheless true that big-band jazz is still with us. In fact, the 1980s have seen it flourishing to a degree that has not been the case since its heyday.

Not only do we have the "ghost" bands whose respective lineages date from the Big Band Era—the Ellington, Basie, Herman, Dorsey, Miller, James, and Buddy Rich bands—we have also those individuals who came up in an early time and remain, consistently or sporadically, active as big band leaders today. There are, for example, trumpeter Maynard Ferguson, who tours worldwide with a scaled down big band and occasionally fronts a full-sized one; Dizzy Gillespie, who in 1987 put together a big band to play the charts of his 1940s unit; drummer Panama Francis, whose Savoy Sultans have re-created the house band of the famous Harlem dance hall; and Artie Shaw, who does not perform with his new band but was its initiator and continues to advise it. Others from the earlier period who have remained occasionally active as big-band leaders are trumpeter Ray Anthony, saxophonists Les Brown, Illinois Jacquet, and Gerry Mulligan, and drummers J. C. Heard (who died n 1988), and Louie Bellson. Others will no doubt come to mind, for we have cited only some of the most prominent of those units or individuals who came up in the Big Band Era and remained active into the late '80s.

Citing a mere several of the exciting contemporary-style units that are extending the big band tradition here and abroad in the late 1980s, we can recommend the eighteen-member New York–based Julius Hemphill Big Band, the Vinny Golia Large Ensemble of Los Angeles, and the international George Gruntz Concert Jazz Band, whose leader lives in his native Switzerland. The repertoires of these three outstanding contemporary big bands are largely by the hands of the respective leaders, alto and soprano saxophonist Hemphill, multi-reed playing Golia, and pianist Gruntz. We emphasize that these are only three notable examples of the continuing tradition of big-band jazz. Several dozen more could be named without effort.

If any bandleader epitomizes the continuum of the big band tradition from the 1930s to the 1980s, it is certainly pianist, composer, and educator Jaki Byard.

"My mother used to give me seventy-five cents to go see the bands that were playing at Quinsigamond Lake—ten cents for the streetcar each way, fifty cents to get into the dance, five cents for a coke," Jaki Byard told me as we sat over coffee at the dining-room table of his home in Queens. He held a Camel in one hand and in the other a pencil, with which he doodled on a yellow pad. A baby grand filled one end of the room. Photographs and memorabilia were here and there: Byard and Ellington together in the early 1970s, Byard with Charles Mingus and Eric Dolphy in Oslo in 1960 or so, and with Stan Kenton later in that decade, a certificate of appreciation from the Rotary Club of Japan, his 1973 Duke Ellington Fellowship from Harvard University, a plaque citing honorary citizenship of the city of New Orleans.

Continuing on the theme of growing up during the 1930s in Boston, Jaki explained, "I would walk to the dance so that I could drink five cokes. I'd stand in front of the band all night and listen. Fats Waller, Lucky Millinder, Chick Webb with Ella Fitzgerald, the Benny Goodman Quartet with Teddy Wilson, Lionel Hampton, and Gene Krupa. That would be about 1936. And I was tuning in on the radio to the broadcasts of the big bands from hotels, 11:30 p.m. to 2:00 a.m. Ellington, Basie, Fatha Hines, Jimmie Lunceford, Benny Carter. Those were the things that inspired me. I guess it stuck with me."

Although Jaki Byard came up during the Big Band Era, his pianistic vocabulary displays a fluency with the entire history of the music from ragtime, blues, and stride through swing and bop to cool and free. He views the music's development as an evolutionary process, a continuum from its earliest years.

Piano was Jaki's first instrument (he began lessons on it at the age of six) and it remains his principal one. Of the other instruments he mastered along the way, Jaki was still performing in the 1980s on alto and tenor saxophones, but he long ago gave up playing the trumpet, trombone, guitar, bass, and drums professionally, although he was still teaching all of them and for compositional purposes is thoroughly familiar with the nomenclature of all of the instruments in the orchestra.

Not long before I interviewed Jaki he had formed his big band the Apollo Stompers. In fact, he put together two versions of the Stompers, one made up of New York musicians and one of students at the New England Conservatory of Music, where he had been teaching for a decade. The student band first came into being as an adjunct of his professorial role and soon was working a weekly gig. But one band wasn't enough for Jaki.

"I was running up and down the road between Boston and New York and I said, 'Why not get a band together in New York, too?' And that's what he did, forming a big band of New York–based professional musicians. A second album of this band was scheduled for release in 1990.

It wasn't long before Jaki had the experience of combining two bands made up of his students in concert at the New England Conservatory, one on each side of the stage.

"I called it the Stereophonic Ensemble," he joked, "and the effect was very interesting because I could bring that band down and this one up and you could hear the difference—just like listening to a stereophonic performance. That was one of my dreams."

Orchestral design for Jaki Byard means "all the possibilities of music—organized sounds, improvisation, freedom." Some of Jaki's musicians have been with free players such as pianist-leaders Sun Ra and Cecil Taylor, and "all of a sudden," he told me, "they have to play free, all just go crazy, and there's nothing I can do about it, until finally, after about five or six minutes, I put my foot down and become the director, 'Either stop this chaos or—!' " He laughed at the thought of it.

I asked him if this sort of "chaos" erupted often. "Oh, inevitably," he responded. "It's a situation that's there. I say, they're gettin' off, let 'em go. Afterward, everybody seems happy about it." Jaki paused at this point, then went on with conviction in his voice: "But someone has to control that type of freedom, there has to be a common denominator, even in a smaller group of, say, five or six musicians. To me, any organization is controlled by a certain person, so in a sense it's a contradiction to say that they have complete freedom in music, although some groups today do have that philosophy that they just get on the stand and start playing and that's it."

In addition to performing the leader's contemporary compositions, Jaki Byard's Apollo Stompers also render tunes of Eubie Blake, Fats Waller, Duke Ellington, and Charles Mingus and include in the program dedication pieces to trumpeters Dizzy Gillespie, Clifford Brown, Lee Morgan, and Woody Shaw and to saxophonists Charlie Parker, Sonny Rollins, and Eric Dolphy. Band vocalists sing ballads, spirituals, gospel, blues, and scat, and "Take the 'A' Train" features a tap dancer. The leader will rise to his feet from the piano bench occasionally and take a few choruses on saxophone. At a club engagement he will stroll back to the bar while doing this.

"My grandmother used to play piano in a silent movie house," Jaki reminisced. "In fact, the piano I first studied on had been given to her by them when the talkies come in. My mother played, my father played, my uncles played—that was the thing then. Instead of the hi-fi set, people played music. If you didn't play, you had a player piano or a crystal set."

Are big bands coming back? Some say that they never went away, and in a sense they never did, for they are still with us, albeit in nothing approaching the numbers of half a century ago. Barney Kessel pointed out to me in 1978 the "sheer economic impossibility" of a return of the big bands, wryly conceding, "The big bands will come back when the fifteen-cent stamp we have now has gone back down to three cents."

As little likelihood as there is that the big band scene of yesteryear will reassert itself as *the* popular musical form of the land, the idiom seemed to have made something of a comeback in the 1980s. As veteran jazz broadcaster Felix Grant observed to me, "People ask, 'When are the big bands coming back,' and I say, 'They're all over the place, they're right here at home—they just don't travel any more.' " Indeed, except for the mere handful of "ghost" bands named above, big bands are no longer found on the road. They might do occasional gigs in another city, and a few of the best make appearances at festivals here and abroad, but it is only those with, for example, the name Basie, Ellington, or Shaw that can command a fee commensurate with the stupendous expenses such travel entails.

That the big band has played a major role in jazz from the early years of the idiom's history to the present is beyond dispute. Indeed, no period of the music's development has been without its large ensembles, and in the case of each of these periods the big band has reflected the essential character of that particular stage of jazz. This can be seen in the theater orchestras of 1920s Chicago, the territory dance bands of the '30s, the big bands of the Swing Era, the bebop bands of Dizzy Gillespie, Woody Herman, and others in the '50s, and the large free-form units that have existed since the '60s.

Moreover, the jazz fan of today can enjoy live performances of all of these styles, for meticulously researched re-creations of everything from 1920s pit bands to 1950s bebop big bands were being presented throughout the 1980s, several historic jazz bands were still active, and many big bands of a thoroughly contemporary nature had established themselves in the major cities and on the festival circuit.

For the big band fan, or music lover, who doesn't often travel, likely enough a big band is based nearby and local concert halls probably book one or another of the touring big bands. So, whether your taste is for one of the older styles or for contemporary expression, make it a point to check out a big band jazz concert, dance, or club date and enjoy a still flourishing art form that is more American than the proverbial apple pie.

6

California

*All we wanted to do was find out how to
play what we were hearing.*
                    Charlie Haden

*It's a feedback music, you have the feedback
from one another, and that's what makes the
music inspired.*
                    Arthur Blythe

*You've got to bet it all, like a gambler.
Everything I have, I've bet it.*
                    Frank Morgan

If New York has for half a century been the jazz mecca, Los Angeles
must today be regarded as the music's most flourishing province. One
close observer of the jazz scene there counted, in the late 1980s, some
eighty active jazz venues in the Los Angeles area, conceding that this
total included piano bars, weekend-only live music, and some groups
of questionable jazz credentials, and asked if New York could top this.

As is the case with the other jazz centers, both principal and secondary, since the idiom's beginnings, it would take a large volume, perhaps several, to adequately document the locale's jazz history. For our
purposes here, the stories of a few representative artists will suffice to
provide the outlines, and supply some pertinent details, vis-à-vis the
jazz scene in Los Angeles since the 1940s. In one case we shall briefly
visit San Francisco.

But first let us point out that Los Angeles, and other parts of California, attracted jazz musicians during earlier periods of the music's
history. The New Orleans bassist Bill Johnson relocated to Los Angeles about 1909 and in 1914 invited the great cornet player Freddie

Keppard and several other New Orleans musicians to join him, although they soon left to tour nationwide as the Original Creole Orchestra. Pianist and bandleader Jelly Roll Morton was intermittently a resident of the city from 1917 for half a dozen years. New Orleans trombonist Kid Ory moved to L.A. in 1919 and stayed for five years; after taking part in historical recording sessions with Armstrong and Morton in Chicago and gigging with King Oliver and others in that city and in New York, he settled in L.A. in 1930 and remained there until the mid-'60s when he moved to Hawaii.

The full roster of musicians, both native Californians and those from elsewhere, who have played roles in the six decades and more of the Los Angeles jazz story would run to many pages. It would include, for example, Mutt Carey, Barney Bigard, Lionel Hampton, Charles Mingus, Benny Carter, Gerald Wilson, Buddy Collette, Gerry Mulligan, Shelly Manne, Art Pepper, Eric Dolphy, Shorty Rogers, Howard Rumsey, Chico Hamilton, Ornette Coleman, John Carter, Bobby Bradford, Chick Corea, Tom Scott, and Larry Carlton. This abbreviated list, in addition to those alluded to in the preceding paragraph along with those whom we shall hear from during the course of this chapter, serves as a mere sampling of the artists whose base has been Los Angeles and whose musical styles have, respectively, encompassed traditional, swing, bebop, cool, free, and the developments that have taken place during the 1980s, both acoustic and electronic.

A player who is not often thought of as part of the California scene is the tenor saxophone giant Dexter Gordon. I interviewed Dexter when he came to the Showboat in Silver Spring, Maryland, for a weeklong engagement shortly after he had relocated in this country in 1977 after fifteen years' residence in Copenhagen. The engagement was his first in the Washington, D.C., area for many years and there was such a crush of press awaiting him after the opening set on that first night that Dexter held a press conference. The appointment that I had made with him to tape an interview for radio was postponed until the following day. We got together in his hotel room for a wide-ranging discussion that took up much of the afternoon, some parts of which were used in our chapter on New York.

"I was born in Los Angeles at an early age," was Dexter's way of beginning his life story for me. "My father was a doctor, a physician. He was a graduate of Howard University's medical school and, funny thing, he played clarinet and mandolin and this helped him work his way through school," Dexter continued, obviously delighted at making this Washington connection, "and he used to play in the old Howard Theatre sometimes, which I did also later on.

"Ever since I can remember, I've been interested in music, in jazz, long before I started playing. I always reacted to it, my mother told me. My father used to take me to the theaters when the big bands

Left: *Frank Morgan (Photo by R. Andrew Lepley)*

Right: *Emily Remler (Photo by R. Andrew Lepley)*

Vinny Golia from the "Large Ensemble", 1989
(Photo by Larry Svirchev)

Big Joe Turner (Photo by Gerard Futrick)

*Paul Acket, The Hague,
Netherlands
(Photo by Rico D'Rozario)*

Bobby McFerrin, 1986 (Photo by W. Patrick Hinely Work/Play)

Ellyn Rucker, Houston, Texas, 1989 (Photo by Tad Hershorn)

Willie Humphrey, 1973 (Photo by Michael Wilderman)

Wild Bill Davison and Edmond Hall at Eddie Condon's Club, New York, 1951
(The Charles Peterson Collection, courtesy of Don Peterson)

George Gruntz and W. Royal Stokes,
Jazzfest Berlin, 1987
(Photo by W. Patrick Hinely Work/Play)

Pete Christlieb, 1989, rehearsing with
the Doc Severinsen Band
(Photo by Paula Ross)

Betty Carter, 1981 (Photo by W. Patrick Hinely Work/Play)

Lester Bowie, Warsaw Jazz Jamboree, 1984
(Photo by W. Patrick Hinely Work/Play)

Dizzy Gillespie and Jon Faddis, 1984
(Photo by W. Patrick Hinely Work/Play)

Pat Metheny, 1989 (Photo by Ken Franckling)

Carmen McRae (Photo by Hyou Vielz)

Author with Bill Cosby at Bill Cosby Studios,
Queens, New York, 1990
(Photo by Gene Martin)

Dexter Gordon, 1980
(Photo by W. Patrick Hinely Work/Play)

*Earl Hines at a private party of Daisy Decker's, New York, November 8, 1950
(The Charles Peterson Collection, courtesy of Don Peterson)*

Don Cherry and Jim Pepper, Jazzfest Berlin, 1986
(Photo by W. Patrick Hinely Work/Play)

Maxine Sullivan holding a thread,
New York, 1937
(The Charles Peterson Collection, courtesy of Don Peterson)

Bob Wilber, June 1949, warming up for Circle Sound recording session, New York. Henry Goodwin, Sidney Bechet, and Bob Wilber (The Charles Peterson Collection, courtesy of Don Peterson)

Dodo Marmarosa, piano, Roy Eldridge, trumpet, Artie Shaw, clarinet, Barney Kessel, guitar, Morris Rayman bass, Strand Theater, New York, 1945 (Photo by Leo Arsene, courtesy of Frank Driggs Collection)

*Bud Freeman and Duke Ellington at a private party, New York, August 1939
(The Charles Peterson Collection, courtesy of Don Peterson)*

Hampton and Basie Sidemen Jam Session, Chicago, 1941

Standing left to right: *Unidentified, Karl George, Ernie Royal, Buddy Tate, Lionel Hampton, Charlie Carpenter, Jack Trainor, Jack McVea*

Seated left to right and foreground: *Harry Edison, Marshall Royal, Irving Ashby, Evelyn Myers, Don Byas, Illinois Jacquet, and Shadow Wilson (Photo by Ray Rising, courtesy of Frank Driggs Collection)* would come to town, which wasn't every day 'cause Los Angeles was quite a hop in those days, you know, five or six days on the train.

"Anyway, my father loved music and used to take me to hear Duke and Jimmie Lunceford and, yeah, Ethel Waters—he was a big Ethel Waters fan, that was his big love—who at that time was married to Eddie Mallory, who had the band, the Mills Brothers, Fatha Hines, Louis Armstrong, all these people, and many of them were, on occasion, patients of my father because, speaking of the early '30s now, I think my father was one of the first black doctors out there. It was a rather small community so, like I say, many musicians had occasion to call for my father's services. So I met many people when I was a kid, people like Lionel Hampton, Marshall Royal, Duke, Barney Bigard, Ivie Anderson, people like that.

"I just always loved music and I always listened to music a lot. When I was about nine or ten I started buying records, building a collection. And how I was able to do that was because there were two or three jukebox operators in my neighborhood who had a garage full of records which you could buy for a nickel, ten cents, fifteen cents, you know, and many of them were jazz records that they didn't use on the jukeboxes. So I was able to go to my little lawn cutting money or paper route money and I could buy up a lot of things. I remember buying a lot of things with Roy Eldridge, his small group, and of course Duke and Lunceford and also Basie. That's when Basie started recording, about '36, which was the year that I first started playing. I was thirteen at the time and my father came home one day with a clarinet. Unfortunately, two months later—I think it was Christmas Eve—he died, had a heart attack. So he never heard me, he never heard me play."

Dexter Gordon's musical associates during his teen years included Buddy Collette, Charles Mingus, trombonist and arranger Melba Liston, and saxophonist Vi Redd. He dropped out of school in 1940, took up tenor, and began playing with a local band, the Harlem Collegians. Late that year he was made an offer he couldn't refuse. He was seventeen years old.

"When I joined Lionel Hampton it was his first band. He had just left Benny Goodman, playing with Teddy Wilson and Gene Krupa and Harry James and Charlie Christian and Cootie Williams and all them. So when he left Benny Goodman he had a very big name and he got this band together in California and, I don't know for what strange reason, he called on me to join the band. Because, really, this was a giant step for me because I was working, you know, around town, but it was mostly on the school level. I didn't think it was anything special, but somehow he called on me for the gig. So I said yes and I said, 'Mom, I gotta go!'

"So anyway, playing with people like Marshall Royal, who I think is one of the world's greatest lead saxophonists, playing with him, and being told what to do and getting punched in the ribs every night, you know, hold those notes out and don't use any vibrato here—playing in a section, huh? Which is the only way you learn how to do it, by doing it and being taught. So, I mean, all this has been invaluable to me as a musician and as a saxophonist."

Dexter spent three years touring with the Hampton band and the several years following that intense experience found him in the big bands of Fletcher Henderson, Louis Armstrong, and Billy Eckstine. (The small combo would be his favored, and virtually only, format after the 1940s.)

Although he returned to California for stays of a few weeks or months during the mid-'40s or so, by the end of the decade he had settled in New York. Dexter starred in the 1986 film 'Round Midnight, and he remained active in the late '80s despite declining health.

"The beautiful thing to me about jazz is that it's such a living music," Dexter reflected. "It grows, expands, broadens, and deepens. There's no end to it. It goes on and on and on." Dexter Gordon died in 1990.

Kansas City native Melba Liston came to Los Angeles in 1937 at the age of eleven. She had been playing trombone for two or three years already. Largely self-taught up to that point, having discarded her first teacher because she felt he was misleading her ("Don't give that man no money 'cause he's not teaching me right," she recalls telling her mother.), Melba began her formal musical training in earnest under Alma Hightower, something of a legendary figure in the history of jazz in Los Angeles.

"I met all the musicians in junior high school who were inclined to pop and jazz and it turned out that there was this lady, Miss Hightower, who had been with the WPA program," Melba recalled for me in an interview at the Maryland Inn, Annapolis, on a morning leading up to rehearsal by her all-female septet. (The group included saxophonist Erica Lindsay, trombonist Janice Robinson, bassist and vocalist Carline Ray, and drummer Dottie Dodgion.) "After the program closed she stayed on at one of the park playgrounds as manager of equipment, you know, bats and balls and things like that. She was trying to encourage all the children of musical talent. We did some little exercises, some one-and-two, and formed a band. She had done this many times before and my age group included her adopted daughter and Vi Redd, who was her great-niece or something like that, and some other nice fellows who went on to become professionals.

"We used to hang out at the playground and play and go in and play music and just live there whenever it was possible. She really started us. She made arrangements and she had us counting scales and all that. There was another band across town so we developed a competition, you know, all guys over there and our band had four girls and about eight or nine young boys. We started playing the stock arrangements, we played music from everything the stocks provided.

"We played at street corners and market openings and the Y, we played at the Golden Gate Exposition in '39, and things like that. Miss Hightower was teaching us show business as well as music. We had to sing, we had to dance, we played harmonicas, we recited poetry, and we did minstrel-type things. She came from a long history of show business, really terrific. We were with her until I finished high school and joined the union.

"Miss Hightower was very angry and hurt when I joined the union. But by then I was ready to go, yeah, joined the union, auditioned for a job at the Lincoln Theatre, passed the audition, and that was about '43. I played in the pit band of a vaudeville house and movie theater. We had the comedians Pigmeat Markham and Dusty Fletcher and the band was led by Bardu Ali, who used to be the front man for the Chick Webb band and could dance a little. We stayed there about two and a half years at the theater and all the acts that would come to town, big-name acts, they'd get the Orpheum or the Paramount in downtown L.A. But every time a black act came along, they not only worked in downtown, they'd come over on the East Side and play our theater.

"It was nice. I started right in doing arrangements. For the show acts we had a big band, about sixteen pieces. And acts would come in that didn't have the proper music to fit our band and the conductor would tell me to straighten 'em out, give me an extra ten dollars. Basically, when we'd do our band numbers we were playing the music from the Chick Webb library, and then we began to get more modern things like Lunceford. We would play some classical things, not too tough, mostly on the comedians' sketches, but not seriously. I mean, we never seriously played a total classical show of any kind. We had this very fine actor who was a close friend of Pigmeat's and Dusty's. He would do a skit, a serious skit, like Shakespeare or something, and we would play some mood music, maybe use a theme from classical, not play the true classical arrangement. And then Pigmeat would come back and do the same thing in slapstick and we would juke it up. That was nice. I got a chance to experiment, do experimental arrangements around those things.

"After they discontinued the shows at the theater, meanwhile, there was Gerald Wilson come to town, who was a trumpeter and arranger, ex–Jimmie Lunceford, very fine, and he kind of lured the younger ones of us from the pit band, stole us from Bardu, actually, and started his own big band. That was about '45. So we started working in my first nightclub, started working Shep's Playhouse, and started going on the road and getting stranded and all those kinds of things. I used to copy for him and I learned a lot studying his arrangements. And

he in turn introduced me to people he was writing for. He'd write for
Basie, he'd write for Duke sometimes, he knew Dizzy. He'd get around
and he was very kind and helped me with my writing. See, if a musician came to town, say, Benny Carter, he had me at their sessions
whenever there was an opportunity to sub for somebody or someone
needed a copyist or anything, Gerald would recommend me. So he
really kinda started me off in the professional world."

Over the course of a long career Melba Liston has contributed to
the books of Billie Holiday, Ellington, Basie, Gillespie, Quincy Jones,
Randy Weston, Tony Bennett, Diana Ross, and Eddie Fisher, to name
only a few; has been a studio musician and arranger for the Blue
Note and Riverside labels; has played and recorded in bands big and
small; and has led her own combos here and abroad. In addition to
her fine musicianship and pre-eminent abilities as arranger, Melba
has served as a role model for many younger women in the art form,
including saxophonists Willene Barton and Fostina Dixon, organist
Gloria Coleman, and, of course, those who have worked in her group
Melba Liston & Company.

The pianist Horace Tapscott arrived in Los Angeles in the mid-
1940s at about the same age as Melba Liston had in the 1930s. "Spirituals and blues, that was about it," he recalled, summing up his exposure to music during his first decade of life in Houston, Texas. "I
was always around music, my mother being a musician and all. There
was a piano right at the front door and my mom put me right on it
when I was about nine." Among the players he heard during those
early years were fellow native Texans the pianists Floyd Dixon and
Amos Milburn. At the age of nine Horace also took up trombone and
baritone horn. He was eleven when his family moved to Los Angeles.

"It was rich," Horace says of the L.A. jazz scene in the 1940s. "That
was during the time they had the black musicians' union, Local 767,
and my residence was a few steps away. I had to pass it each day so I
got to know most of the musicians. Gerald Wilson was the first man
that pulled me off the streets. I was about thirteen and he took me
in. Later he put me in his band. Everybody there who was of any kind
of knowledge was always passing it on to me. There was a constant
flow of knowledge from the pros.

"I was down the street from all the nightclubs. They had clubs up
and down Central Avenue next to each other and we lived on Central
Avenue in the midst of everything. Each way I went I could hear
some live music and I got the chance to listen to them rehearsing
down at the union. They had three floors so three different groups
would be rehearsing and you'd get to hear them all. When I first
heard Hamp's band I said, 'Wow! I'd like to get in this band.'" Horace recalled hearing the bands of Fletcher Henderson, Ellington, and
Basie along with the combos of Charlie Parker, Gillespie, and Big Jay

McNeely. When he was about sixteen he saw Art Tatum crossing the street one day. "I was too young to go to a lot of gigs," Horace chuckled, "but I was tall, so I slipped in." He was playing in local combos from the age of fifteen.

Trombone was Horace's principal instrument when he began playing in school bands in his early teens, and his fellow band members included trumpeter Don Cherry. At sixteen he was asked to join Gerald Wilson's band, and he remained in it for a year or so. In the mid-'50s, while a member of an air force band, he took up the piano in earnest, and at the end of the decade he went on the road with Lionel Hampton, thereby fulfilling an early ambition.

Upon coming off the road in 1961 Horace formed the Pan Afrikan Peoples Arkestra, of which he became leader and chief arranger. It is interesting that the advanced musical ideas that characterized this unit mirror in some respects the emerging developments in jazz that were occurring in other locales during the early 1960s, most notably in Chicago with the formation, in 1965, of the Association for the Advancement of Creative Musicians (AACM). The Ark—as it soon came to be known—had a community-oriented organizational wing, the Union of God's Musicians and Artists Ascension (UGMAA). The Ark and UGMAA gradually became the focus of much community activity, attracting youngsters from Watts to participate in workshops and join the band. UGMAA produced concerts at churches and schools, the parks, and on the street corner, and eventually documented the sounds of the orchestra in a number of albums.

"To begin with," Horace explained, "it was a thing about preserving the music and strengthening the musicians in their craft, that was the idea. A lot of these guys were dying at the time and I had a lot of talks with them during my coming-up days. They put a lot of bugs in my ears about different things, you see, and you start looking at things around you and adding them up and putting them together. The only way to do it is to do it and there was a lot of things that I didn't like that was going on with the musicians. So I just figured that if we could just start building our own kind of group and have some kind of standards to work from, then maybe we could receive some kind of respect for what it was that we were offering. I thought perhaps the Arkestra would be the ideal kind of thing.

"The Arkestra was made mostly of the kind of musicians that most conventional bands didn't want to deal with. So we put them all together. It was to form a unity and preserve something, Royal, and to have people learn to respect this music that comes out of Afro-America. The reason we did this was we figured that people weren't getting exposed to it. So the Arkestra was the only way we knew of that would be some kind of tool to use to expose the music to the people. In other words, we took the music to the people by way of an orchestra with all of its formalities to try to get them more used to what was going on and so they could learn to support it. That was the general reason for having the Arkestra.

"It was unique in its way because it had its own kind of coloring about it and its own standards about it and people started hooking up to it, knowing that it was music coming out of the community. A lot of folks used it as their launching pad, speakers and everything else, because the music was a kind of building of a pride within oneself, the way we used to put it over. We would play the music in churches and on school grounds and on the street corners, just exactly where the people were that we were trying to meet, trying to get to. We had to bring it to them, and after a while of bringing the music to them, they wanted to know where the music was.

"It took us a long time, man, it was years, I'd say a good fifteen years before they even started thinking about us as working. There was no money there in the black community and we were on our own for a lot of things. But it helped build up some kind of courage and pride. It took us about fifteen years before we could even think about anything, monetarily speaking. A lot of things had to happen, like working on the side—in my case I was ghost-writing for different people just so I could be where I wanted to be—as well as raise your family.

"It took us quite a while before they recognized us as a group, before they would accept us, but through the years we've always had a regular place to go every day, every night. We would get billed sometimes in old homes [that] people would let us have and we would rejuvenate them and call them UGMAA House and we would use that for our rehearsal. So we had no problem with facilities through the years. . . . We always had a place to go. The whole idea was to stay together, to get over before people started recognizing us, you know, wanting us. So we went through three generations, that Arkestra, and today there are a lot of people in positions here in L.A. that are returning favors and memories and different things. But it's not to . . . the halfway point, to where we want to get it. It has a start and it's still working and the guys are still rehearsing on the weekends in the high school. So it's still active and it's still the goal to reach seventy-five pieces by the end of the century. Most of the concerts now are outside at a lot of festivals, the Watts Music Festival three times a year, and at museums around town and at churches and in the parks.

"They were standing in line." Horace said of youngsters from Watts wanting to become associated with UGMAA. "It wasn't just the instrumental playing; it had to do with whatever their talents were that had to do with the art. We were there to encourage it and we had someone there to help them in it. A lot of them are still playing. There are some that are in high school and college now and still playing and some have grown up and are still playing. There's been maybe, out of ten, two that stay with it." Saxophonists Azar Lawrence, Jesse Sharpe, and Arthur Blythe are three artists whose early experience included membership in the Pan Afrikan Peoples Arkestra.

A musician whose network of associations has extended into many areas of jazz and who settled in Los Angeles in 1956, bassist Charlie Haden grew up in Shenandoah, Iowa.

"I started singing on the radio with my family when I was two years old," Charlie told me. "We had a radio show every morning and every afternoon. Then we moved to Springfield, Missouri, and we were on the radio there until '49 or '50, and then we went to Omaha, Nebraska, and we had a television show there when TV was first happening. The show stopped when I was fifteen. We were singing like folk songs and what they called hillbilly music and mountain music back then, you know, what they call country-western now, and then we sang some religious songs like 'Will the Circle Be Unbroken' and 'Mansion on the Hill.'

"All my brothers and sisters played guitar and my older brother played bass," Charlie explained, "and I was listening to classical music and jazz on the radio. And then my brother brought home some records of Jazz at the Philharmonic with Bird and Dizzy Gillespie's big band and Billie Holiday. It was very difficult to get records there in the Midwest. I really liked the music, really thought it was beautiful, and I started, whenever I could, picking up my brother's bass and playing with the record. I was about fourteen. I was completely self-taught."

After finishing high school Charlie turned down the offer of a full scholarship at Oberlin College in order to attend Westlake College of Modern Music in Los Angeles, arriving in the city at the age of nineteen.

"I started thinking about playing and wanting to learn more about music and I arrived in L.A. in 1956 and started immediately playing with different people at night from Westlake and then meeting other musicians and going to jam sessions, and I dropped out of school. I went and sat in with Art Pepper's band and he hired me to play the next weekend with him, and the first night Hampton Hawes was playing piano, the next night Sonny Clark was playing."

Summarizing the next couple of years, Charlie continued, "I was playing a lot of different gigs and I met Paul Bley at a jam session and he hired me to play with his quartet with Lenny McBrowne on drums and Dave Pike on vibraphone. We played at this club called the Hillcrest Club. While all this was going on I was going to a lot of jam sessions and playing and becoming kind of impatient because I was sometimes hearing—" (in his head, that is) "—ways of playing, of improvising, not on chord structure but on the feeling of a piece. Sometimes I would play like that and try to fulfill what I was hearing and other musicians would become very upset and they wouldn't know where to come back in after my solo. I'd have to play the melody of the song, you know, to get them to come back in.

"So one night I went to this place—I think it was called The Haig, a jazz club—and this guy came in to ask if he could play and he got up on the stage and that's the way he was playing, on the feeling of the piece or the inspiration of the composition rather than on the chord structure. And the guys that were playing at the time asked him to stop playing immediately.

"I ran backstage to try to meet the guy and he was gone. And the next night at the Hillcrest I told Lenny McBrowne that I heard this alto saxophone player with this brilliant sound and ideas playing. And he said, 'Was he playing a plastic alto?' I said, 'Yeah.' He said, 'That was Ornette Coleman.' I said, 'I sure would like to talk to him,' and so Lenny said, 'I'll invite him to come and meet you.'

"So Ornette came by at a Sunday morning jazz session and Lenny introduced us," Charlie recalled from three decades before our interview. "I told him, 'You played beautiful, man,' and he said, 'Well, I don't hear that very often. Thank you. Would you want to come over to my place and play some music after you finish here?' I said, 'Yeah.'

"So we went over to his house and I was finally able to fulfill the different feelings that I had about improvisation other than the chord structure, 'cause he was doing that, Ornette, that's the way he played, and it was like all of a sudden starting on a new venture or something. We started rehearsing over at Don Cherry's. I already knew Don and Billy Higgins. We had played gigs together. . . . I was still at the Hillcrest and then in 1959 Don and Ornette went to the Lenox School of Jazz in Massachusetts for one summer. John Lewis heard them and told Nesuhi Ertegun about them, and that's when Ertegun came out to L.A. and came over to a rehearsal to hear us and then we made the first record, 'The Shape of Jazz to Come,' for Atlantic Records in L.A. Then we went to New York.

"During this time in L.A. . . . Paul Bley was in the hospital with an ulcer thing and Elmo Hope was out here and replaced him for a couple of weeks at the Hillcrest and, boy, was that a treat, 'cause Elmo Hope was like a master. And all during this period Scott LaFaro was staying at my apartment and Scott and I became very close friends. He came running in the apartment one day with this record and he said, 'Man, you've got to hear this piano player! I gotta play with him some day.' It was Bill Evans' first trio record.

"Then Paul hired Don and Ornette and Billy at the Hillcrest Club and kept me there and he let Dave Pike and Lenny go and we started playing as a quintet. Then the business started falling off. We had a big clientele built up with Paul's quartet with Dave and Lenny and we had stayed there playing every night for about a year. And then Ornette and Don and Billy came on the gig and it was just beyond the average listener that came in there. The audience gradually thinned out to just musicians would come to hear us, and right before we left for New York the gig ended.

"It was really exciting," Charlie enthused about his first trip to the big city. "I always thought about going to New York and I always wanted to go to New York 'cause most of the music that I listened to was coming from New York and I had met a lot of musicians in L.A. that lived in New York. So arriving in New York was really exciting. I'd never seen anything like New York. I guess one's first experience in New York, first day in New York is kind of—you're in awe.

"We went over to the Five Spot and we had a rehearsal. I remember when we came up to the Five Spot to rehearse there was this wino that was lying on the sidewalk in front of the door. We couldn't hardly get the door open, you know, and Lou Donaldson was coming out of the club as we were going in. I guess he had been rehearsing there with his quartet. And I reached down to help this guy, you know, 'cause he was out and I thought he was hurt. Lou Donaldson said, 'What're you doing?!' I said, 'I just want to help this guy up.' He said, 'Don't bother him. He's happy.'

"Opening night was unbelievable. I mean, the club was packed with people, mostly musicians. One of my students at California Institute of the Arts, where I head the jazz studies, once asked me, 'Why do you always close your eyes when you're playing?' Well, naturally the reason for that is concentration and listening, but I told him [this] story. One night early on in our stay at the Five Spot I was unpacking my bass and getting ready to play the first set and I looked out and standing at the bar was Charles Mingus, Wilbur Ware, Paul Chambers, Ron Carter, Richard Davis, Percy Heath, Henry Grimes—I mean, every great bass player in jazz was standing at the bar looking dead in my face, waiting to see what I was going to play. From that time I closed my eyes when I play!" Charlie burst out laughing at the memory of this intimidating scene.

"We caused quite a commotion. People used to argue, scream at each other," Charlie continued, recalling those weeks in late 1959 at the Five Spot Cafe. "Fisticuffs broke out. It was exciting. And we were just thinking about the music. All we wanted to do was find out how to play what we were hearing, because it was like a new language, man, but we didn't think about it as a new language, we just thought about it as how we heard this way of playing and how we wanted to do it. That's what we were all thinking about, we weren't thinking about what people were writing or what people were saying, we were just thinking about playing, and we would have rehearsals and we would play and we would talk about what we were doing.

"It was a very exciting time, kind of like, I don't know, maybe a similar feeling when Bird and Diz were getting together, and maybe a similar feeling in New Orleans when jazz was being born. Because, you know, if you think about art forms, the beginning of an art form or the beginning of an addition to the vocabulary of the art form, it's kind of awe-inspiring, like when, in New Orleans, they had no one else to listen to, they were their own influence, or like the first people who made film, who did *they* have to look to?

"That's the way it was playing with Ornette," said Charlie. "You were creating something for the first time. It was like an urgency, a desperate urgency for spontaneity that's unlike any other improvising that you'll ever do."

Charlie Haden found himself once again in the very same format in 1987 when Ornette Coleman re-formed the quartet, bringing together Don Cherry, Billy Higgins, and the bassist for concerts and festivals in the U.S. and overseas. I caught them in October of that year at Jazzfest Berlin and was impressed with the combo's high level of creativity and the interplay between the members, which seemed little short of ESP. Speaking briefly with both the leader and Charlie after the set, I could not help but conclude that both still shared that sense of urgency.

In addition to his role as educator, Charlie was, in the late 1980s, traveling for months on end worldwide in a variety of contexts, including collaborations with pianist Geri Allen and drummer Paul Motian, keyboard player Patrice Rushen and drummer Terri Lyne Carrington, saxophonist Joe Henderson and pianist Kenny Drew, pianist Ran Blake, bassist Egberto Gismonti, and his own Quartet West. The 1989 Montreal Jazz Festival honored the bassist with *Les Sessions Charlie Haden: Un Hommage,* a feature in which he performed on each of the event's nine evenings with different musicians, including some of those named above as well as guitarist Pat Metheny, pianist Paul Bley, and the Charlie Haden Liberation Music Orchestra.

"We did the first record in 1969, *Liberation Music Orchestra,* on Impulse," Charlie responded to my request that he provide me a capsule history of his Liberation Music Orchestra project, the twentieth anniversary of which was upon us. "Then in 1982 we did another record on ECM called *Ballad of the Fallen.* The first thing that inspired me was the music of the Spanish Civil War that I'd heard and the second was the direction that the country was going politically.

"In 1969 Vietnam was at its height, Cambodia had just been bombed by Nixon, Che Guevara had just been murdered in Bolivia by the CIA. I have always had a feeling of being concerned about equality, being from the Midwest, a very racist part of the country. There's a lot of *great* things in the Midwest, there's a lot of beautiful things in the Midwest, but there's some ugly ones, one of them being racism, and I saw that and the aspect of organized religion, all the poor people giving their money to a guy that has a mansion and Cadillacs, and at a very young age I saw all this, blacks being discriminated against. And I've always had this feeling of wanting to have brilliance in the world, to have the leaders of the government want to *protect* life rather than destroy it, to emphasize the brilliance of life and to put all their energy and resources into *enhancing* life instead of destruction and greed.

"And so I wanted to make this record and just communicate my feelings. I felt very lucky to have the medium of music in order to communicate this. I think that the art form of jazz is a political art form anyway, really. I mean, I think that just the fact of struggling to bring very deep music to people all over the world is a struggle in itself, and also to communicate human values to people is a political commitment. I mean, that's talking about equality and making the world a better place and I think that whether musicians or artists realize that or not, I think that's the end result of what they're doing.

"So I just wanted to kind of bring it to fruition and make a political statement and that's when I made the first album. That was music from the Spanish Civil War and a song that I composed for Che Guevara called 'Song for Che' and a song that I composed about the Democratic convention in Chicago and then a piece about Vietnam and a lot of different things of Ornette's. There was 'War Orphans' of Ornette's. The album won the Grand Prix in France, it won the Best Album of the Year gold award in Japan and Best Album of the Year in Melody Maker, and it was nominated for a Grammy. It won a lot of awards but they didn't promote it because they didn't really like the politics of it," Charlie said, adding that the musicians in the orchestra for the first recording included Don Cherry, drummer Paul Motian, trombonist Roswell Rudd, clarinetist Perry Robinson, guitarist Sam Brown, saxophonists Gato Barbieri and Dewey Redman, tuba player Howard Johnson, trumpeter Mike Mantler, and pianist Carla Bley, who did the arrangements.

"The second album was made after Reagan was elected and I saw the country going back in the same direction again and I felt the need to make another political statement. This was when El Salvador was in the news so we did this poem called 'The Ballad of the Fallen.' On the back of the LP is a painting by a child who was a refugee from El Salvador who was living in Nicaragua.

"And now we're going to do a new recording. There's music from South Africa, the African National Congress anthem, a piece that I wrote for a documentary film that I scored about Nicaragua called 'Fire From the Mountain,' a spiritual I've written for Martin Luther King and Malcom X that's going to be sung by the Harlem Boys' Choir, two pieces from Cuba written by Sylvia Rodriguez, a composition by Carla Bley that she's written to a poem by Langston Hughes called 'Dream Keeper,' a song from Venezuela, a song from El Salvador, and another song from the Spanish civil war. Carla's written the arrangements and it's going to be really, really beautiful."

Arthur Blythe, a Los Angeles native who spent his school years in San Diego and returned to L.A. in 1960, told me that he began taking an interest in music from an early age and was soon "listening to blues music, blues records my mother had around the house. Wasn't all blues, though, because she used to listen to Earl Bostic and Duke Ellington and Tab Smith. I remember those people. I happened to be listening to some Texas blues one time and I got an inspiration at that time—it had been being created all along—and I begged and pleaded with my mother to get me a horn and she did, a Christmas present when I was eight or nine, a saxophone, an alto, and that's when I begun."

Arthur "immediately got a couple of little sensible notes out of it and that was really inspiring and encouraging." He soon found that he could reproduce the sounds he heard on records. "That was one of the qualities that gave me encouragement to continue." A year or two later he was playing in the school band, and he was playing in local combos from the age of thirteen. Upon returning to Los Angeles when he was twenty he began a long association with Horace Tapscott, becoming a founder member of UGMAA and the Pan Afrikan Peoples Arkestra.

"I was growing as a person and as a musician," Arthur pointed out, "but it wasn't like me hanging out with name people and like I was a prodigy of the heavy music scene. It was just people trying to do what *you're* trying to do, too." As to the similarities between what the Arkestra and UGMAA were doing, musically and in terms of contribution to the community, and what the Chicago-based AACM was doing during the 1960s, Arthur insisted, "It was just coincidence. We had never met them. In retrospect, I see that the sound of the music and some of the ideas that they had about betterment of self and raising the level of consciousness in the community to be better Americans and better people, to be better for themselves, to raise our whole level of consciousness—those were also our base aspirations and we went about them in our various ways. And, just looking back, I see that they were associated."

One of the groups Arthur Blythe has led in recent years is called In the Tradition and it is aptly named since it draws on the early jazz tradition for inspiration and includes in its repertoire compositions by Fats Waller, Ellington, and other masters of the early decades of the idiom. "I was raised in the tradition," Arthur explained. "That's one of my points of stress, that's what I like, to come from that perspective, because that's what the music is about from my point of view. Whatever you deal with, it's about tradition now. It's like making cornbread without cornmeal—it becomes another type of bread then.

"Yes, I was raised in the tradition. It was like my environment, my surroundings. It's another thing to research the history of that tradition, and some of that research turns up some of the feelings that you had experienced coming up, too. It might have been applied in a different way but you find the associations, the hook-ups, the connections, you know. I've looked at the tradition as closely as I could and can because I am a part of that. It is more or less like a way of life, being involved in a way of life."

Citing an aspect of that tradition that has played a significant role in his musical conception, Arthur referred to the collective improvisation characteristic of the early New Orleans style. "I think that's essential, that's important to me, especially how they put that together. That's how I feel about it when I be making the music, to play in a collective sort of way. A rhythm section and a soloist, that's less personal. The music, historically and traditionally, was dealt with at different times from a collective point of view. Maybe all the way through it has to be collective because it's . . . feedback music, you have the feedback from one another, the guys who are making the music, and that's what makes the music inspired. . . . I just prefer the collective. I think, also, that's a part of the heritage."

Arthur Blythe relocated in New York in the mid-'70s, as we indicated several chapters ago, and he has both led his own groups and worked under the leadership of Gil Evans, Lester Bowie, and others. Since 1984 he has been a member of The Leaders, the other members of which are Bowie, Chico Freeman, Kirk Lightsey, Cecil McBee, and Don Moye. "It was like education to me," Arthur observed, thinking back on his involvement with various groups and bands. "I gained things from working with Chico Hamilton and Gil Evans, Horace Tapscott, Jack DeJohnette, Lester Bowie. There are a lot of people you can get knowledge and information from about what's happening. I mean, you might have missed some details on the way and someone can always show you something to help the picture come together clear. It's reciprocated, too, you know—I show them certain things as they show me. It helps everybody, raises the levels of everybody. And those qualities I gained from these people, I gained from other people, too.

"It's always been a struggle," Arthur reflected, musing on the future of the music, "and I see the struggle as not lessening at this point. But I don't see it as deterring my efforts any. I'm going to stay positive and I am staying positive so that I can keep on going on."

Another player who came up in California and later settled in New York is the trumpeter and Dizzy Gillespie disciple Jon Faddis. Jon grew up in Oakland, where he took up trumpet at the age of eight, inspired by seeing Louis Armstrong on the Ed Sullivan show. Jon told me that Armstrong "broke it up, he tore it up on the show." It wasn't much later that Jon's parents decided that they would like the youngster to study music and asked what his choice of instrument was. "Remembering Louis Armstrong, I said, 'Trumpet!' And the next day I'm going down on Saturday afternoon for trumpet lessons."

When he was eleven Jon was turned on to jazz by his teacher Bill Catalano, a former Stan Kenton trumpeter and an ardent admirer of

Gillespie. "He told me, 'Dizzy Gillespie's the greatest trumpet player living.' I got a little Dizzy 45 that I memorized and I started fanatically buying all of Dizzy's records. The stuff he plays! I'm still trying to get it! In high school I started developing my technique a little more and my upper register a *lot* more. I didn't really get into improvisation until I was fourteen or fifteen."

Jon played with local r&b combos and sat in with San Francisco big bands through high school during the late 1960s. His first exposure to his idol was at a Basin Street West performance in San Francisco when he was thirteen. "I was too scared to go up to him and say, 'Hello, I'm a trumpet player' and 'Could I get your autograph?' When he came off the bandstand and walked past me he patted me on the back—I still remember that."

Two years later it was the fledgling trumpet player's mother's turn to escort him to a Gillespie session, this time at the Monterey Jazz Festival. "I took about fifty of Dizzy's albums," Jon remembered with a laugh. "We saw him walking where all the booths where they sell T-shirts and stuff are and I said to my mom, 'There he is! Go get the albums!' So she ran and got the albums for me and he sat down on the grass and autographed every one of them. I still have them.

"A few weeks later I went to see him at the Jazz Workshop, where no minors are allowed, but my mother took me and they let me in. I had my trumpet with me and Dizzy was playing 'Night in Tunisia' and I asked him if he was going to play the ending and he said, '*You* do it, you got your horn.' So I ran downstairs to get my horn out of this dressing room and came back up and he was signaling with his eyebrows and I missed the signal. There was this big pause and he took his horn away from his lips and I looked up and played the ending and everybody in the club turned around—" (Jon was standing in the rear of the room) "—and he invited me up to the bandstand to play and made the statement that I was fifteen years old but had been playing sixteen years. We played a couple of tunes, 'Satin Doll' and 'Get That Money Blues.' On 'Get That Money Blues' I played some of Dizzy's solo and I think that surprised him."

Upon graduation from high school in 1971 Jon joined the Lionel Hampton band for six months. Settling into the New York Jazz scene at the age of eighteen, he became a member of the Thad Jones–Mel Lewis band for three years of Monday nights at the Village Vanguard and a tour of Europe, the Soviet Union, and Japan.

Jon Faddis seemed to flash across the horizon for the three years or so after he arrived in New York. He made appearances at the Newport in New York festivals and Radio City Music Hall, went on a European tour with Charles Mingus's group, concertized with Sarah Vaughan in Carnegie Hall, was invited twice to Dick Gibson's Annual Colorado Jazz Party, and performed on the European festival circuit.

Jon even filled in for an ailing Roy Eldridge at a Mingus date and he turned up on albums with Gillespie and Oscar Peterson.

Placing high in the polls during this period, Jon was in first place three years running in *down beat*'s Talent Deserving Wider Recognition trumpet category in the magazine's International Critics Poll. Charles Mingus opined that Jon had the ability "to be one of the greatest trumpet players in the world," and Thad Jones said that he was the best since Gillespie. Dizzy himself heaped such praise upon the barely-out-of-his-teens trumpeter that, in the view of many, the line of succession seemed all but established. Then Jon allowed himself to be swallowed up by the studios, unable (by his own admission to me) to reconcile his lack of confidence in himself with the glowing praise of fellow musicians, the press attention, and the pressure of the public life.

From the mid-1970s until the early '80s Jon concentrated his attention on studio work, applying his virtuosity to pop materials, backing up Frank Sinatra, Billy Joel, Stanley Clarke, the Rolling Stones, Kool and the Gang, and many others. There were jingles, the soundtrack for the Clint Eastwood film *The Gauntlet*, the theme for *The Bill Cosby Show*, even a commercial for Japanese whiskey that Jon and Clark Terry collaborated on.

The degree to which the technology has taken over in the studios disturbed Jon. Describing a session for a George Benson album, Jon recalled how the horn sections were dispersed here and there, the rhythm players relegated to still another corner of the room, and the musicians isolated under headphones. "What happens," he lamented, "is that the engineer starts to control the music and the musicians really lose control of what's happening to their music. I don't know, I think it could have been done better. It gets a little frustrating in the studios. It's almost as if anyone can go in and do it who has any measure of competence on the trumpet. Financially it was great. But it's not the kind of playing that's good for the soul."

The turning point for Jon came in 1982 when Dizzy selected him to accompany him in an In Performance ceremony at the White House in which several major American artists showcased young colleagues whom they believed to be "on the verge of exceptional careers." With Chick Corea, Miroslav Vitous, and Roy Haynes providing the rhythm support, Gillespie, his protégé Jon Faddis, and Stan Getz rendered "Groovin' High." Then, joined by vocalist Diane Schurr (Getz's guest) and violinist Itzhak Perlman, who emceed the event, they jammed on "Summertime."

While Jon has some serious misgivings about their hosts' clumsy handling of the ceremony ("They were asking Stan Getz about bebop and Dizzy was standing two feet away"), there is little question that the occasion provided the impetus that the then twenty-nine-year-old trumpeter needed to convince him that he had much, much more to offer the world than horn section lead on jingles or the odd solo, often uncredited, on a lavish studio production of some pop star.

It wasn't long after that pivotal event that Jon decided that he "wanted to get out and play more," which is precisely what he commenced to do and has continued to do through the 1980s, leading his own groups for club dates, appearing on the festival circuit with the Dizzy Gilles-pie Big Band, and recording in various jazz formats.

"I'd like to get rid of the image of being a studio musician and a protégé," Jon told me soon after he had emerged from his period of self-imposed confinement in the studios. "I'd like to establish myself as an artist in my own right and do music that people can identify with, music that is accessible not only to the jazz fan but to people that might just walk in off the street. I think sometimes musicians get a little too cerebral."

Much seemed to have come about by the late '80s, with one con-spicuous exception, the anticipated erasure of the tag "protégé" whenever the name Jon Faddis comes up. Indeed, Jon may well have to live with that for a long time to come—not exactly a cross to bear, when you think of it.

It was on the occasion of a weekend engagement with his newly formed quintet at Washington, D.C.'s One Step Down that I observed the combo in action. The small club was packed and the leader's as-tonishing velocity and glass-shattering highs on the opening "Unlim-ited" left one keenly anticipating what was in the offing. It was crystal clear that this was to be a class act. Nor was it simply flash, for Jon was harnessing his extraordinary technique to the expression of emo-tional nuances, a truth that his blazing "Struttin' With Some Barbe-cue" bore witness to. "Whisper Not," a Benny Golson tune, had Jon's horn close to the mike with mute, pinching and bending notes, and "Early Bird" featured a boppish torrent from his trumpet.

The combo was nicely balanced between Jon's tilt to the past and alto saxophonist Greg Osby's contemporary expression, a complemen-tary relationship not only recognized by Jon but agreeable to him. He remarked to me the following day, "Greg will play more modern—that's his style." James Williams also contributed mightily to the eve-ning's program, demonstrating that he is as much in the tradition of the great masters of melodic jazz piano as he is thoroughly modern. His solo on the Armstrong classic was a gem.

The virtual withdrawal of Jon Faddis from jazz performance for those years spent in New York studios calls to mind a similar experi-ence of a gifted player and composer of Philadelphia origins who, before settling in Los Angeles in the 1960s, had spent time in the groups of Gillespie and Art Blakey, co-led a sextet with Art Farmer, and written a number of jazz standards. We are speaking here of tenor saxophonist Benny Golson, whose compositional skills were employed by the television and movie studios of Hollywood. For a decade and a half Golson dropped out of sight and sound, as far as the jazz audience was concerned.

"When I first went to Hollywood," Benny told me upon his reappearance in the mid-1980s, "I was trying to live down the jazz label, the bebop kind of thing, because with that I would have been put into a corner as to what I could do. 'Oh, he does jazz stuff.' It helped me to broaden myself, as far as writing for film, so that I could write comedy and dramatic music, different situations, period music, things like that, so that I had no limitations. So it was worth it, for the while that I did it, backing off a bit, because when I went there it was constantly, 'Who is Benny Golson? What has he done?' Because I wasn't in the jazz idiom per se, I was in another idiom, in an idiom where I wasn't really known, so I had to gain credibility, I had to gain some credentials, which I was able to do.

"The remuneration really was better," Benny admitted, "staying there doing the writing, but sometimes you felt like you wanted to be released from something, even for a moment or two. Sitting there at that desk, day in and day out, like a hunchback, you know, writing, writing, writing, writing, making deadlines. The family goes out to Malibu Beach on a sunny day and I walk to the front door and that's as far as I can go and I go back into the dungeon. I didn't play the horn for seven or eight years, which presented a problem when I got ready to come back. It was like getting over a stroke. I never liked stopping playing, but I just thought it was over when I went out there. I even sold about half my instruments and I guess I would have sold them all, but there was something inside that said, 'Well, just hold on to these for a while.'

"I did *M*A*S*H* for three years, *Mission Impossible, Mannix, The Partridge Family,* a black version of *The Waltons* which hasn't been aired yet. I did television and radio commercials, too."

Then in 1982 a call came from a producer in Europe who wanted Benny to reactivate the Jazztet, the group that he and flugelhorn player Art Farmer co-led in the early 1960s. Benny contacted Art, who had settled in Vienna, Austria, in 1968, and trombonist Curtis Fuller, a Jazztet alumnus. Both agreed to come back together in the old format, and the combo was re-formed with Mickey Tucker at the piano, Ray Drummond on bass, and Marvin "Smitty" Smith on drums.

Speaking of the years he spent at his desk grinding out background music for film and TV, Benny insisted, "My mind wasn't completely idle. I was *thinking* playing jazz, I was *thinking* writing jazz when I wasn't doing dramatic music for TV and whatnot. Now I'm going back and the things that I was just thinking about are some of the things that I'm putting pen to paper now. In the last three years I've composed more things than I composed in the first part of my life." Benny's return to playing was the result of more than simply the need to re-establish direct contact with the jazz audience, a relationship that is to a considerable degree the raison d'être for the jazz artist; it was also inspired by "a feeling like you want to be released from something. And I'm enjoying it, I'm just delighted, very tickled about it. It's like therapy for me."

The other side of the Hollywood connection for a jazz player is the studio. Pianist, composer, and arranger Bill Mays, who has recorded and performed with Sarah Vaughan, Gerry Mulligan, Golson, and many others, worked in the Hollywood studios from the late 1960s until relocating in New York in 1984. A California native, Bill's film score credits include *Superman,* the television series *The Fall Guy,* and others which he could not recall, he confessed to me. "My cloudy memory is a statement on that business, on the people that work as musicians in that business. You're on call through contractors who call orchestras together for certain composers. If you're busy, you typically do ten or twelve recording dates a week, each one for a different picture, commercial, or television show. It gets to be a blur after a while."

For bassist Red Mitchell, a veteran of countless big bands and combos since the 1940s and former first bass at MGM, the last straw was of a philosophical nature. "I didn't want to continue contributing to the atmosphere of violence there, doing television and movie music which I felt was contributing to the violence here in this country. I guess I have to take my share of the blame," he admitted. "I was playing on the *Peter Gunn* show in the late '50s, which was one of the first of the shoot-em-ups that used jazz as background music." Red moved to Stockholm, Sweden, in 1986, but makes periodic visits to the U.S., lending his virtuosity to various combinations of first-rank jazz artists like himself.

Tenor saxophonist and Los Angeles native Pete Christlieb has been, since the mid-1970s, one of a chosen few to enjoy the best of two worlds, a steady job as a member of the Tonight Show Band on Johnny Carson's nightly program and gigs on the side in L.A. clubs with Louie Bellson, the late Frank Rosolino and Warne Marsh, and visiting stars the likes of Mel Tormé.

"I don't know how much attention the television audience pays to the Tonight Show Band," lamented Pete, "but people who like the music all have the some opinion, 'Why in the hell don't they let the band play more?' The function of the band is to keep the audience— who come from Montana or Iowa or took a bus from Arizona or wherever to see the Johnny Carson show—from looking at each other and saying, 'We could have stayed home and watched commercials!' The show is a hundred thousand dollars a minute for commercials and they show a lot of commercials. The band is there to keep people occupied during those commercials, keep them from sitting there gawking at television monitors hung all over the studio. It's the excitement of the music live that keeps things happening in that studio. I do my best work behind dog food commercials."

It is the exposure Pete has had as a featured soloist on the Carson show that has brought him other work in the studios of Los Angeles. Over the years Pete has contributed his sound to the *Sonny and Cher Show, Star Trek, MacGyver, Dynasty,* and other series, but he says that "Hollywood is not Hollywood any more" and that he wouldn't be getting the few contracts that still come his way were it not for the network of friends he has for years enjoyed the loyalty of, like the arranger Dennis McCarthy, without whose help, he says, "I wouldn't have any studio work."

If anyone should know whether anything of the old Hollywood, vis-à-vis the working musician, remains in place in the 1980s, it would be Christlieb, whose father, an oboe player, worked in the major film studios, including MGM and Twentieth Century-Fox, for three decades.

"He did *Gone With the Wind* and worked with Igor Stravinsky," said Pete proudly. "Stravinsky was in the house rehearsing with my father all the time. I would go and watch my father make motion pictures. I was taking violin lessons and following in his footsteps. I first heard jazz when I was twelve or thirteen and I fell in love with it. Saxophone players like Zoot Sims and Gerry Mulligan and people like Chet Baker. There was the feel of it all that made me really want to play jazz. What it meant was that you could decide where you wanted to go musically. You didn't have to sit and hack away at something, like repetition."

Pete's schooling, both high school and a semester at Valley College, included band training, and he went on the road at the age of eighteen with the Si Zentner band. "It was a learn-while-you-earn situation," he recalled, adding that for the remainder of the 1960s, and into the early '70s he spent a year with the Woody Herman band, backed vocalist Della Reese and others in Las Vegas shows, and went on the road with Louie Bellson and Pearl Bailey. For several years in the early '70s Pete was periodically a member of a traveling version of the Johnny Carson Show, and when it relocated in Los Angeles in 1976 he was asked by Doc Severinsen to join the studio orchestra.

"The business is almost extinct here," Pete observed of the contribution of musicians to the soundtracks of film and television productions. "There is no longer an orchestra per se in a Hollywood studio like there was in those days when they did motion pictures every day practically. A lot of it has been taken over by synthesizers and a lot of it has gone out of the country. Some of our well-known composers here are working in Yugoslavia, Hungary, England, where music and labor are cheap. That started in my father's era. They began to do that in the '60s when musicians were striking for better wages and a cost-of-living increase. When they needed to get a score done they'd do it in England while there was a strike going on here. Eventually, some of the great motion-picture success stories like *Star Wars,* the whole thing was done by the London Philharmonic!"

"I wanted to be in studio *recording* and not stealing microphones," alto saxophonist Frank Morgan told me at lunch in a Washington, D.C., restaurant. It was his first visit to the city and he was in town for a weekend gig at the One Step Down. We had met for the first time a year earlier at the North Sea Jazz Festival in Holland and I had recommended the club as an appropriate venue for him. "I made a lot of money in recording studios stealing the very microphones that I demand to play into right now, the best. They cost twenty-five hundred, three thousand dollars. Two in each sock. I'm just saying that I've come a long way. Because it was a dream of mine to one day come back in these studios and not have to steal anything. I don't play when I steal. That's why my career was—is—like it is. I didn't have time. Bebop don't pay enough money. I was serious. I used fifteen, sixteen hundred dollars every day, average. Let's just say I was a very good criminal. I mean, I did it with gusto. I really went at it.

"I never had any [arrests for] possessions," Frank said, pointing out that his convictions were always for theft. "The closest I came to having possession was when they busted me for tracks, and being under the influence. They would start out calling that 'internal possession' or something, but as far as them catching me with the stuff, no. I've gotten my ass beat 'cause I wouldn't give the stuff up, but I didn't throw it. I fought until I swallowed it—and then got indignant." Frank laughed at the recollection of such a scene from the past.

"Obviously I was a real actor. I played many roles. I played Joe Convict, Benny Dopefiend, Charlie Parker—I played the shit out of Bird! I tried to out-Bird Bird. 'He's got it!' That was my motto. 'You got it, Bird, you got the whole thing!' I was always playing somebody else. In fact, probably the first time I played myself was on stage," Frank suggested, referring to his featured role in the 1987 off-Broadway semi-autobiographical production *Prison-Made Tuxedos,* written and co-directed by George Trow. "I'm right at home on the stage. I don't know if I can play anybody but myself, but there's many selves."

Frank Morgan relocated to New York in 1988, having visited the city for the first time only two years earlier, but his story for most of his life before that move is a California-based one. It is not a happy story for the three decades or so after Frank began serving the first of his numerous prison terms.

"I went to jail the first time in my life in 1953, San Francisco." (He was nineteen years old.) "Kenny Drew and I got busted together. I was working with Oscar Pettiford and he was with Buddy DeFranco, at the same club. And I was in and out of jail since 1953. I mean, in and out and in and in and in and in and out until November 8, 1985.

That's when I was released two weeks after *Easy Living*"—Frank's first album in thirty years—"was released. That was my last week of prison, which was different than any other trip, not just because I had a record coming out, 'cause I'd done that before, but because I turned myself in. I went back to prison the last time 'cause they were looking for me when I did *Easy Living*. When I finished *Easy Living* I knew I didn't want to go to prison any more. I wanted to play bebop.

"My history of heroin addiction was from 1952 until the present," explained Frank, who has been on a methadone maintenance program since his final release from prison in 1985. "Because once you're an addict, you're always an addict. I haven't found a way to come off methadone without going back to the streets. I hope to come off, but I would rather stay on methadone for the rest of my life," Frank insisted, recalling the many companions of his youth who had been destroyed by the drug. One of them, "a helluva alto player, went crazy. Well, I did, too, but he went certified. His was civil, mine was criminal. That saved my life, incidentally, the California prison system. It was very kind to me in that respect, but in other respects very unkind to me. They almost made me feel like I wasn't in prison. They let me play all the time. I even had bands that went on the road from prison." In fact, Frank was co-leader of the San Quentin big band, a showpiece for visitors, along with fellow altoist Art Pepper.

"I was born December 23, 1933, in Minneapolis," Frank began, responding to my request that he tell me about his early years. "I don't know exactly what age I started playing guitar but my mother told me recently that my father"—the late Stanley Morgan, who played in the bands of Harlan Leonard and others and with original members of the Ink Spots Charlie Fuqua and Deek Watson—" would have her hold his guitar, when she was real big carrying me, and he would go around behind her and reach around her big belly and play the guitar. So the back of the guitar was against her stomach and the sounds were going right into me. When I came out he would sit by my crib and play the guitar eight, nine hours a day, when he was there, and when I started reaching for the guitar that's when he started teaching me. So I don't really know how old I was. His dream was for me to be a great guitarist.

"At six I moved to Milwaukee to my father's mother, see, 'cause my mother and father were on the road all the time. So my grandmothers, maternal and paternal, raised me. At seven or eight my father took me over to Detroit for Easter vacation, something like that. He took me to hear Jay McShann's band at the Paradise Theatre and Bird stood up to play his solo on 'Moody Blues' or 'Confessin' the Blues,' one of the two, and my father says I jumped up and said, 'That's it! I want to play that!' He took me backstage and I met Bird." (Stanley Morgan and Parker had been fellow members of Harlan Leonard and His Rockets out of Kansas City.) "And Bird was sup-

posed to meet me the next day, meet my father and I, and pick out
what I thought was to be a saxophone. It turned out to be a clarinet.
Bird didn't show up. Wardell Gray and Teddy Edwards, who were
both playing alto in Howard McGhee's band, went with us.

"So I played clarinet first, for maybe a year or two, and then I went
to soprano, then to alto. A lot of guys had sopranos, I think. That
was the case with my teacher, a tenor player named Leonard Gates.
He had a soprano, but he never played it, he played his tenor. I was
ashamed of my soprano because it was straight and the case looked
funny. So I used to sneak through the alley to my lessons until I got
an alto. It was just to hold over until my folks could afford to get me
an alto, when I was about nine or ten." Frank pinpoints his increasing
awareness of alto saxophone greats as taking shape at about this age.
He was listening to, among others, Johnny Hodges, Benny Carter,
Louis Jordan ("bad cat"), and, of course, Parker, with close attention,
on record, on radio, and, not much later, in person. After all, his
father had settled in Los Angeles by the mid-1940s and opened an
after-hours club, the Casablanca, where many bebop players of note
routinely performed, including Parker during his California period
(late 1945 until the spring of 1947).

"I went to Los Angeles the first time at thirteen, spent the summer
of '47, went back to Milwaukee," where Frank was living with his paternal grandmother. "The next summer my father was getting ready
to go out on the road with a band and my grandmother found a big
marijuana joint in my shirt pocket. She called my father in California,
told him, 'I think it's time you had your son. I just found a joint in
his pocket,' she said. 'It wasn't one of mine—it's better than mine.'
That's what she said. She took me to Chicago and put me on the train
to California, got me a compartment and paid off all the porters and
everything to watch me. When they were getting ready to pull out of
the station she backed up against the door of the compartment and
locked it and went in her bra and gave me a supply of joints to last
the trip." Frank chuckled over the memory. "She's dead. That's why
I play 'Love Story' for her. I talked about her in the play. When I was
in San Quentin she was on her deathbed and it was her last wish that
she could see me. I didn't get back to Milwaukee until '86 when I went
to a family reunion. My father came home from Hawaii, where he
was still playing, in Honolulu. I went to my grandmother's grave and
played 'Love Story.'

"At fourteen years old I was about to graduate from high school,"
said Frank, returning to the story of his teen years. "I was in the
twelfth grade, 'cause in Milwaukee they had something similar to the
Montessori method. You could go at your own pace. If you could
prove that you could do the work at another grade, they would skip
you to that grade in Milwaukee. But when I got to California I found
out that they didn't play that, they didn't do that, they didn't graduate

people early, particularly black people. My transcripts and tests were at college level but they held me up until I was seventeen. Actually, I regressed, I think, academically."

In Milwaukee, Frank went to Lincoln High School with the pianist Willie Pickens and the saxophonist Bunky Green. The peer jazz connection continued, in fact accelerated, upon his arrival in Los Angeles.

"I lived initially on Main Street on the borderline between east and west Los Angeles. I lied about my address—we used some friend's address—so that I wouldn't have to go to Jefferson High School. You see, my people didn't want me to go to Jefferson because they wanted me to get an education, so I went to Manual Arts High School. After they wouldn't let me in my proper grade and after I found out they were playing bebop at Jefferson, there was no stoppin' me. I wanted to go where they were playing music and so I transferred and went to the all-black Jefferson High School so that I could be under Dr. Samuel Brown, a pianist who played big fat changes. There was a full curriculum of harmony and counterpoint and the whole afternoon was big band."

This was the band, Frank told me in an aside, that had produced Dexter Gordon and Art Farmer earlier and would, several years later, have in its ranks Don Cherry and Billy Higgins. "Horace Tapscott was a trombone soloist in the band. Also, he was a helluva scat singer. All kinds of guys. There was an alto player, a guy named George Newman, that to this day I think played better than Ornette. I can't think of any other big names that came out of the band. Some cats went other ways, like rhythm and blues.

"I'm talking about a *band*," Frank emphasizes. "I mean, this was 1948, '49, and we were playing the same shit that Dizzy's band was playing and all the Woody Herman stuff—'Lemon Drop' and 'Manteca'—plus Dr. Brown was writing and all the cats were writing. It was a bebop band and we were playing concerts all the time, all over the state of California, we were on television, and we were in all kinds of contests. We were sought after in all the schools and colleges and we were only in school two or three days a week. It was really a great training ground. Plus I had a band outside of school."

For a man in his mid-fifties who had spent something like half his life behind bars on burglary convictions, Frank Morgan appeared to be a remarkably well-adjusted individual, especially in view of his "habit," which he had come to terms with only a few years before the afternoon that we sat talking at lunch. On several subsequent occasions Frank seemed, if anything, even more relaxed. For all of his glowing accounts of how splendidly prison had treated him, it was clear that here was a man who had been through hell several times over.

"You see, I spent my whole life either living in the past or dreaming of an unreal future," Frank explained. "This is why the play as written will probably never go to Broadway, because I wouldn't want to play it every night, not the San Quentin part of my life. I've played that. I thought I was a hot-shot thief and dope fiend. I spent all my life in prison and I know that that ain't shit. I had to come full circle to understand the desire to fail. I *know* about failure. I'm ba-a-d-d when it comes to failure. You dig?

"The most important thing that you can tell about me in prison is that I did it until *I* got tired. So when I quit, it ain't no problem because I want to succeed and *I* made that determination. *I* chose. I said, 'Well, when I get tired of doing this shit, if I still have some life left in me, then I want to try success. I don't know how good I am at success, but I'm sure going to find out.' You've got to bet it all. It's what I do now, man, like a gambler. Everything I have, I've bet it. Little old me—who else could I bet? If you can't bet on yourself, what the fuck is it good for, this life? It looks like I can do success better than I could failure, because I got hooked. Life has taught me—I find out now—that the hippest thing in the world I can say is, 'I don't know. Show me. Help me, I'm trying, I'm teachable.'

"So I'm seventeen all over again. I'm a bebop saxophonist, that's what I do, and I'm doing it with gusto. I want to play, because I like to talk shit. I'm not puttin' people down, but you ain't heard nobody talk shit," Frank insisted, asserting that what he has to say on his horn is the real thing, the stuff that jazz expression must convey if it is to be meaningful, that is, the player's life experience.

"I'm just saying that I got tired of being a damn fool. That gets old. So it ain't no contest, you dig? I *love* success, I love sitting here talking with you in Washington, D.C., talking about the last time I saw you, and it ain't telling no lie, it ain't no penitentiary trip. This *here, this* is what's happening."

A number of jazz artists have expressed themselves in the graphic or plastic arts. A by-no-means comprehensive roster of performers who have done so include: Sonny Rollins, Meredith d'Ambrosio, Les McCann, Pee Wee Russell, Marian McPartland, Bob Haggart, Mel Powell, Tony Bennett, John Heard, and George Wettling. Miles Davis has had shows of his sketches, George Gershwin painted portraits, and Dee Bell has been active in both hand-modeled and monumental welded sculpture. The painter Larry Rivers, who played saxophone in the bands of Herbie Fields and others (and was still playing, in the 1980s, with other artists on a part-time basis in the 13th Street Band), turned to painting full time in the 1950s. Vocalist and songwriter Gene Lees has worked as a commercial artist.

For virtually all of these, music came along first. In the case of the Los Angeles–based leader, composer, and multi-reed and woodwind player Vinny Golia, it was the other way around.

"When I was out of art school I started teaching and doing some things in New York, some shows and stuff," Vinny told me. He pointed out that he had grown up in the Bronx and relocated in Los Angeles in 1973. "I worked on rather large canvasses, wall-size—small for me would be about six feet high, maybe seven, eight feet long—and I was finding that visual art was lacking a little bit in immediacy for me."

Vinny was already at that time an avid jazz fan with a large record collection, and he often had an album on the turntable as he painted. "I could actually hear the lines when I was painting," he explained, "and I had certain favorite albums that would always turn up."

Vinny also hung out at jazz clubs and found himself sketching the musicians. "The drawings would take the shape of the way the person soloed. Since I had been schooled pretty well, I had a good grasp of human anatomy and I could always relate it to human form. What I was trying to do was to get the essence of the solo into some kind of form, the player being the actual receptacle." Drummer Tony Williams, pianists Cecil Taylor and McCoy Tyner, saxophonist Ornette Coleman, and the late trumpeter Lee Morgan were among those captured by Vinny's pencil. "The ideas that I had sketched in the club I would use later to paint from."

One thing led to another, and in 1971 Vinny found himself doing a show with saxophonist Dave Liebman's trio. "We had a dancer and a percussionist and I did live paintings, big canvasses, right on the wall. I was really into that." Similar sessions involving pianist Chick Corea, multi-reed player Anthony Braxton, and others followed. Vinny also experimented with musicians playing to his completed paintings. "Each musician was assigned a color and he could follow it through the painting. Since I was into doing a lot of optical illusions, he could start at any point and travel through as his eyes were looking at the color."

Not long after these experiences Vinny moved to California and made the transition to musician. "I made enough money to buy my first saxophone and started to get serious about music." Over the course of the next few years Vinny applied himself to a grueling sixteen-hours-a-day schedule of self-instruction. As a result, he now performs on virtually all of the saxophones, flutes, and clarinets, as well as on some exotic instruments such as bamboo flute.

Vinny Golia has been, since the beginning of the 1980s, at the very center of contemporary West Coast jazz. His associations have included such other leading California musicians as Horace Tapscott, trumpeter Bobby Bradford, clarinetist John Carter, pianist Wayne Peet, and trombonist John Rapson. Vinny has traveled widely, performing at European festivals and in Japan, and in many parts of the U.S., including New York and Washington, D.C. On the road here and abroad Vinny has played with a diverse selection of artists among whom have been the guitarists Sonny Sharrock and Elliot Sharpe, Japanese altoist Akira Sakata, Indian drummer Tricky Sankaran, and German trombonist Conrad Bauer. Vinny was invited in 1989 to join the George Gruntz Concert Jazz Band for its spring tour of Europe and the U.S.

"One thing in our favor," Vinny pointed out, making a connection between his world-wide travels and his self-produced record label, "is that from the *9 Winds* record thing people seem to know who we are." Another thing in his favor is that he is a self-starter type. "Somebody says, 'You guys sound pretty good. We'd like to have you play for us next time you're here,' and we say, 'Sure,' and we take their names and it just gets a little bigger. If you don't take a chance to go any place, you never get exposed and you never get a chance to meet the people that are really behind the music.

"We've been taking our large ensemble music [scores and charts] and bringing maybe three or four people from the ensemble to a specific place and that'll be the core of the thing and the rest of the nineteen players are from there. We get there three or four days ahead of time and we have three rehearsals, an intensive kind of thing, and then we do the performance. I did this in Europe and it works very nicely. It's a chance to not only get the concept of your music around, to spread an idea about the way you approach improvisation and composition and stuff, but to also hear how other people improvise in your context and to expose them to you."

As for pinning a name on the style or idiom that Vinny and his colleagues express themselves in, "It's kind of hard," Vinny admitted, somewhat amused. "I mean, it would be great if we could find some kind of tag to put on it. We have a lot of different elements in it. It used to be that if you played freebop, you never played any changes, but now we play changes, we play ballads and a lot of lyrical things. It's really a collage of angular vamps, different odd tempos—just a kind of contemporary style of music which happens to be uniquely American."

7

Post
Bebop
Developments

*A lot of people think that avant-garde jazz
music came out of the blue, but actually it
came out of the blues.*
                    Dewey Redman

*If I can make as much money as my mail-
man, that's a good year for me.*
                    Lester Bowie

*If you heard it yesterday, we try not to play
that tomorrow.*
                    Sam Rivers

While this chapter will deal with the so-called avant-garde, or "free
jazz," developments of the 1960s and '70s, the reader will note the
conspicuous absence of women in the following pages. This is certainly not because they did not contribute in the area of "free" playing. Rather, we have postponed to the final chapter the stories of a
number of female instrumentalists. Since that chapter deals with the
current scene of the 1980s and because women have begun to come
to the fore this past decade or so as instrumentalists in a way that has
never before occurred in jazz, it seemed that their stories would most
appropriately be contained therein. Singers, male and female, whose
approach has partaken of an avant-garde, or "free," nature, have been
similarly excluded from this chapter and considered instead in the

chapter devoted to singers. Nor will avant-garde players from abroad
be touched upon here but rather included in our chapter on jazz in
other parts of the world. As for the contributions of Horace Tapscott,
Ornette Coleman, and others on the West Coast, they have been covered in the chapter on California.

"I came from the South West blues tradition and I used to go hear
a lot of musicians who came through Fort Worth like T-Bone Walker
and Louis Jordan," saxophonist and native Texan Dewey Redman recalled. "There were no concerts, there were only dances. In every
town they had a place for a dance. But there were also some very
good bebop players in Fort Worth and Dallas. And of course Ornette
[Coleman] grew up in the same environment.

"At that time [the 1940s] it was still very much segregated and all
the black kids in one town had to go to the same school, one black
school, no matter which side of town you lived on. Consequently, I
met Ornette and we played in the same high school band. We had
what they called a 'jump' band, a little swing band, then Ornette started
playing bebop, and then he got into his own thing in Los Angeles. I
never realized that he would later become one of the greatest. But the
seeds were there before he left Texas and to me it was not anything
unnatural to hear him play what was later to be developed as avant-
garde. He was just a local cat who could play, a local guy everybody
liked.

"A lot of people think that avant-garde jazz music came out of the
blue, but actually it came out of the blues. I'm saying that anyone that
came out of the southern background, as opposed to California or the
East, they had to be exposed to the blues because that was a way of
life. It's not remarkable, for instance, that Ornette Coleman came out
of that and yet created something very different.

"Actually, I don't consider myself strictly an avant-garde player because I try to play a variety of music. It's okay if that's what you want
to call me, but I prefer the term 'musician.' "

The foregoing reminiscences and observations of a great artist and
educator serve, better than historical account by the author, to reaf-
firm the evolutionary development of jazz that has been a prominent
theme of this book. The jazz researcher will seek fruitlessly evidence
that the idiom is anything but a continuum from its earliest days to
the present, a truth that Dewey made succinctly and with wry humor.

Dewey Redman's career as performer reaches back four decades to
his participation, while attending high school, in rhythm and blues
groups and swing combos in Fort Worth. He taught in high school
for four years before moving, in the early 1960s, to San Francisco,
where he co-led a big band with saxophonist Monty Waters and worked
with saxophonist Pharoah Sanders, bass and reed player Rafael Garrett, and guitarist Wes Montgomery, among others, and led his own
small groups. In 1967 he relocated to New York and was re-

acquainted with his friend of high school days Ornette Coleman, whose group he became a member of until 1974. From the inception of his New York years until the late 1980s Dewey has performed and recorded with a number of artists and groups, including Charlie Haden's Liberation Music Orchestra, Keith Garrett's group, Carla Bley, Leroy Jenkins, Roswell Rudd, and Pat Metheny. Dewey's own groups in the '70s and '80s have included Old and New Dreams, the other members of which were Don Cherry, Ed Blackwell, and Haden, his former associates in Coleman's quintet.

Tenor saxophonist Pharoah Sanders, with whom Dewey Redman sometimes found himself on the bandstand in San Francisco in the early 1960s, is another whose roots are in the blues. "At first I wanted to be an artist, a painter," Pharoah told me, musing on his high school years in Little Rock, Arkansas, "but I got kind of wrapped up in music and I kept my art secondary."

It's little wonder that the horn won out over the brush, for before he was graduated Pharoah was playing with local blues groups and on one occasion, when he was seventeen, he sat in with B. B. King. Several years later marked the beginning of Pharoah's friendship with John Coltrane.

"When I met him at a club in San Francisco he was having trouble with his mouthpiece so the next day we went to pawn shops and places like that looking for one," Pharoah said, recalling his first encounter with Coltrane in the early 1960s, on the occasion of which they spent the day checking out hundreds of mouthpieces. (Coltrane had for years been seeking the perfect mouthpiece, allegedly once traveling to Florida to have one custom-fashioned from the armor plate of a German tank.)

"In New York I met him again at the Half Note. We used to talk on the telephone about health foods and exercise and he asked me to come down and sit in," Pharoah said, admitting, "I got sort of nervous."

From these casual beginnings blossomed a friendship that eventually led to an artistic collaboration that has been described as cross-fertilization between free-jazz pioneer Coltrane and his disciple Sanders, who was then in his mid-twenties. The association lasted for about two years until Coltrane's death in 1967 at the age of forty. "That was a time when he started hearing some other kinds of music and other energies and he was trying to find new sounds," Pharoah reflected on his mentor's creative vision, which sprang from a spiritual awakening that found expression in the 1964 *A Love Supreme,* an album that sold a quarter-million copies, an astonishing figure for a jazz release.

Pharoah Sanders went on to lead his own groups throughout the 1970s and 80s, documenting on many recordings his signature sound of fierce overblowing and forays into the outback of harmonic territory. His associations have included pianist Alice Coltrane, widow of the saxophonist, vocalist Leone Thomas, pianist John Hicks, drummers Idris Muhammad and Billy Higgins, bassist Ray Drummond, and Don Cherry.

"There was always music around my neighborhood and in my house," violinist and native Chicagoan Leroy Jenkins told me (and my radio show's listeners) in an on-air interview. He named "Charlie Parker, Dizzy Gillespie, and those guys, a lot of blues singers" as the staples of his musical diet as a youngster in the 1940s, adding that he was "raised up with an auntie who was four or five years older [and] listening to all the singers—Billy Eckstine, Louis Jordan."

Leroy took up the violin when he was eight. "My first violin teacher, he wasn't very good violinistically. I mean, he wasn't the one that really put me on my feet violinistically, he just sort of introduced me to music and the power of it. The fact that I took from him for so long and that he was responsible for some of my direction is why I remember him more than I do all my teachers. But, actually, I didn't really start playing, I mean, really getting the technique of the violin, until I met Bruce Hayden at Florida A&M in 1956. In other words, before then I took lessons but I wasn't really very serious with it."

Nevertheless, Leroy did perform on violin at church functions most Sundays from the age of nine until he was fourteen and in high school, at which time he "switched over to clarinet," clarifying that the black schools in Chicago at that time did not have orchestras, only concert bands, and thus there was no opportunity to play the violin.

"I sort of like lost interest in the violin. Any teenager would have because it wasn't the style then. Charlie Parker was the teenage idol at the time. I was going to hear him play a lot, dancing to his music. So that's how I was introduced to the music, really, it was a teenage thing. After a while I started to play little, small-time jobs around Chicago with the saxophone."

Leroy Jenkins continued to play alto saxophone with bebop groups until he left Chicago to attend Florida A&M, and it was at this time that he returned to the study of the violin, having given up the horn. Graduating in 1961 with a degree in music education, Leroy went on to instruct in stringed instruments in the public school system of Mobile, Alabama, for four years.

"I came back to Chicago from Alabama in 1965 looking to play for a living instead of teaching school," Leroy continued. "In the 1960s, when revolution was in the air and freedom was the cry, the musicians, the black musicians, who had been under the heel of the business world for so long, woke up and found out that there was a certain amount of freedom that they had to have to pursue further what they were into, and in order to do that they had to find out more about the business of the music and, number two, try to seek avenues, other avenues, of expression beside the club. And, besides, it was a new type of music that was coming, too, as a result of Ornette Coleman.

"So it was all these things. There had to be some new-type situation for the musicians to exist in, to co-exist in. So I think that, amongst other reasons, that's why the AACM was organized. I'm sure there's other reasons, I'm sure that other people could think of other things, but I would say those are some of the more important ones. The AACM—it means Association for the Advancement of Creative Musicians—was an organization that had its beginnings through the frustrations of musicians.

"Well, when I got back to Chicago, all my contemporaries—John Gilmore, Clifford Jordan—were in New York. I wanted to go to New York but I knew I wasn't ready at the time, musically. The reason why was because I wasn't doing anything new. I was playing the violin and I wasn't doing anything exceptional, so much so that I would stick out. I don't think any of the guys playing that kind of [new] music was looking at me favorably.

"So when I first heard the music of the AACM [which the pianist Muhal Richard Abrams had founded in May of that very year, 1965, of Leroy's return to Chicago], it was something different, something where I could really be violinistic. In other words, I didn't have to worry about staying in any particular key or range of the instrument. To be more specific, I found out that when you are playing tunes, so-called tunes, that were in a certain key and a certain form—either ABA form or AABA form—they all seem to be played in a certain position and the only time you get a chance to play high was when I was reaching for a note, or low as when I was reaching down to a note. I mean, I could never get the two together. Maybe it could have been a lack of facility, could have been that, you know, 'cause, after all, that was ten years ago and even though I played well, there's certain nuances that one must have in order to play great. I was, at that time, at most a good player, and you can't play *well* in this business, you gotta play *great*. So after getting in the AACM and seeing how they operate, I discovered that I would be able to play more of my instrument and I wouldn't have to worry about the clichés, you know, I wouldn't have to play the 'doo-da-lee-doo-da-lee-bops.' I found out that I could really soar, I found out how I could *really* play.

"Sometimes I was amazed at so many great musicians co-existing together. That was one of the unusual things about the AACM. There were around twenty-five, at least, active musicians and more than that inactive. All those names that you always speak of and hear of—Anthony Braxton, Roscoe Mitchell, Muhal Richard Abrams, Steve McCall, Leo Smith, Kalaparush Maurice McIntyre. All these guys, we met and saw each other every week for four years and we played together. I played just about every kind of situation, with just about everybody. I played with the Art Ensemble of Chicago, I got the full advantage of the AACM. I was able to write music, something I hadn't been able to do before, hadn't had the opportunity, hadn't had a vehicle. I *had* a vehicle there, we had a twenty-piece band. So I was able to write some music and bring it and it would be played. That was a big plus because that wasn't happening to me nowhere. It *wouldn't* have happened to me nowhere.

"And at the same time I was playing with guys who were much younger. Both Leo Smith and Anthony Braxton were years younger than I was, and Anthony, at that time, had pretty much a direction, a musical direction. That's something I didn't have then. So I have to say that he helped me a lot in the sense that *he* had a direction. You know, sometimes you're not able, when you're out here reading *down beat* and listening to the interviews, to really see how a direction is come by. I mean, when you're actually in this business, you never see it unless you're dealing with somebody that's actually dealing it. Very seldom are musicians lucky enough to deal with innovators. With the AACM I was dealing with a whole bunch of innovators. I was able to see people who knew what they wanted to do and I heard some stuff that I hadn't heard before. See, they were using their *imaginations,* and that made *my* imagination soar, you know what I mean?

"We all had to go to Europe," Leroy said, jumping ahead to 1969. "That's the only place that we worked. We didn't work in Chicago, that's for sure. Oh, we worked at the Lincoln Center, but these were things that we sponsored and the money was very short. We couldn't live off of it. Our group [the trio of Leroy, Braxton, and Leo Smith was known as the Creative Construction Company] was considered very way out, even more so than the Art Ensemble, so we didn't really get even the acceptance that they got. At that time the Art Ensemble at least had one direction—that was where Roscoe took it—but in our group we had three directions, 'cause I was writing, Anthony was writing, and Leo was writing, and all very powerfully. We were looking for a drummer when we went to Europe and Steve McCall, who had played with us before on different projects in Chicago, had been there for a little while so we got together. Of course he would be the logical choice since he was in the AACM.

"Now practically all these guys in the AACM that I came up with are in New York," Leroy observed of the late 1970s, "and we're on the New York scene and we're doing quite a bit there. I would say that we work more, we have more attention, more people are coming to hear us now. I think that we have developed our art form to a much more together level, I'm not going to say 'understandable' level, because one thing I would like to say is that this music that we are playing is sort of a classical music based upon black music or jazz or however you say it, and this music is not the kind that you are used to hearing in clubs or something like that. We sort of like direct this music toward the concert stage. It's cerebral and the improvisation is now at an extent that it covers the whole orchestra, so to say, like it doesn't stay in one key or one mood or one this or that. I mean, it's trying to create an aura.

"And I think that we've just about got to the point that we're not now playing as much nonsense as we did before, not that we ever *intentionally* played nonsense. I think that, whenever you're starting off on something comparatively new, that there's a certain amount of nonsense that has to be included in order to experiment.

"Like I said, I think that for the most part we now understand what we're doing, and *now* what we're trying to do is get it out there. And so we are playing more, like I'm here in Washington, D.C., for the first time and I'm up here talking on a radio station, something that I do all the time in New York, and I'm getting out there more, playing more places than I've ever played.

"I'm also playing with a quite different instrumentation now, Amina Claudine Myers, piano, Andrew Cyrille, drums, and I am approaching this trio in a different way, in that it's based on improvisation and—how should I say it?—empathy amongst us. I don't require that the pianist or the drummer accompany me. All I ask them to do is add to what I'm doing and to try to make it a mix, a swinging mix, you know, a groovy mix, but not like swinging or grooving in an old way but in another way, the kind where the swing is more of a swing than a"—he tapped on the table top with a steady beat—"than a 'pop.' Of course it's not much different from the old thing, it's still the 'head,' as we say, the motif, and the improvisation, still the same ingredients, but they have been rearranged and the ideas are different, they're based on other things beside form. In other words, our form can be anything, and that includes sixteen, twelve, and thirty-two bars."

Leroy Jenkins has remained on the leading edge of the music into 1990. In a later conversation he spoke of his then new, nearly all-string group Sting (two violins, two guitars, bass, drums) in these terms: "Sting is the culmination of all the music that I've been playing since I started playing and since I started writing. In the process of survival here I have given lectures on jazz, starting with the 1920s and up to the present—Louis Armstrong, Zutty Singleton, Duke Ellington, Charles Parker, John Coltrane—and in the process of doing this it started creeping into my music. With Sting I'm able to use music in the tradition but in sort of a changed around or hybrid form. I want a string kind of jingle, a new sound that I can put this music that I wrote to, since everything I've listened to I've always tried to relate to the violin."

Of all the musicians who have risen out of 1960s and '70s avant-garde obscurity—an obscurity even insofar as the general jazz audience is concerned—the trumpeter and leader Lester Bowie has perhaps had the greatest success in becoming, if not a household name in the jazz community, at the very least a performer with considerable name recognition among those who follow the art form. I first interviewed Lester in a dressing room of the Kennedy Center in Washington, D.C., on the occasion of the Kool Jazz Festival, a noon-to-midnight event. I continued to get together with him whenever he visited the city as a member of the Art Ensemble of Chicago or as leader of one or another of his own groups. Lester's story is one that confirms me in the belief that the jazz artist, regardless of style or period, comes up in analogous circumstances, circumstances that shape his or her art and give it a character that makes it uniquely American.

Lester Bowie was born in Frederick, Maryland, in 1941, spent several years in Little Rock, Arkansas, and grew up, from the age of about three, in St. Louis, Missouri. It is not surprising that a five-year-old Lester was already blowing trumpet, for three brass-playing uncles were members of the Bartonsville (Maryland) Cornet Band around 1911 to 1915, and Lester's father, W. Lester Bowie, had been director of the Dunbar High School Band in Little Rock in the late 1930s. (Photos of both of these bands are reproduced on the cover of Lester's *All the Magic* on the ECM label.) Lester Sr. continued to pursue a music career when the family relocated in St. Louis, as teacher, concert band trumpeter, and bandmaster.

"I took lessons from my father," Lester told me, laughing at the memory, "and it was not by choice, it was by complete force." Describing his father as a "hard taskmaster," Lester recalled daily practice, frequent parades, and hearing music all the time. "These southern bands ran up and down the street and I remember we used to have records around the house—Ellington, Basie, Louis Jordan, Bull Moose Jackson, all those sorts of things."

Before he was very far into his teens Lester was playing for dances, school affairs, parties, "and every now and then we'd play in a joint." He described the music that he and his peers played as "a mixture of r&b, Dixieland, and boogie-woogie, a pretty weird sounding band, I would imagine. We did one thing on tape—I'd like to find it—on a radio show. I'd like to hear it because I can't imagine what that shit sounded like."

Of the many St. Louis trumpet players Lester grew up hearing, he cites in particular Bobby Danzie as an important influence. "He would show me bebop, all those hip tones, and really nice warm sounds like Miles and Art Farmer." Lester adds that it wasn't just trumpet players he listened to. "There were a lot of good horns of all sorts and a lot of *terrible* saxophone players," he insisted, employing the adjective in its complimentary sense, that is, as a higher degree of "bad."

Four years of army life from age seventeen instilled discipline in Lester and taught him organizational skills, and he averred that he "had a ball" the whole time he was in the service. Stationed in Texas, he absorbed the blues the state has long been known for and he was playing in local r&b combos. Upon discharge Lester attended college briefly in St. Louis and then returned to Texas to continue his education. "I didn't go to school, I must have flunked out," he said, convulsed in laughter. "I played with James Clay, Fathead Newman, all the cats from Ray Charles' band—they were all from Dallas—and that's when I really dropped out."

Thus began Lester's first taste of the road, three years of it, with blues and r&b bands the likes of Albert King and James Brown, and even a carnival or two. "Every back road in the United States, buses and cars, years of it. You lived out there. Sometimes it would be two or three days before we stayed in a hotel. Drive five hundred, seven hundred miles, gig, then go change clothes, get back in the bus and move to the next gig before we'd get into a hotel. I lived like that for years."

Marriage to r&b singer Fontella Bass (they have since divorced) and her Top 40 hit single "Rescue Me" combined to bring the trumpet player to Chicago in 1965. "My wife was with Chess Records and we decided to move to Chicago and work out of there for a while. We had been on the road quite a bit and I had been directing her show. When we first got there I was working in other bands, doing a lot of studio work, commercials and things like that. I was there a while before I actually met the members of the AACM.

"I had always wanted to be a jazz player," Lester explained. "I never wanted to be a director of a rock show or anything like that. So I was always looking for that even when I was playing these other type gigs, looking for places to play, looking for cats to play with, looking for musicians to listen to.

"I was getting kind of bored and there was this baritone player named Delbert Hill who was with a lot of the same bands I was with and he said, 'There are some guys you might enjoy playing with,' and he took me to a rehearsal." As soon as Lester walked in the door on that occasion of his first meeting with members of the AACM he knew that he was finally in his element.

"True jazz musicians are eccentric personalities, right? And you're used to seeing a few of these at a time. I had always known a handful of guys like that, guys you run with, guys you get together to play with. But the AACM, when I first met *them*, it was like twenty-five guys in the room and it was the largest amount of really these type musicians that I'd ever seen. And that was quite exciting for me." Lester named off a few of those he met that first night—Abrams, Mitchell, McCall, Kalaparush, Thurman Barker. "By the time I got home from the rehearsal Roscoe was calling me, asking me to join his group.

"There was no work at all and what we were doing, the whole idea of the AACM, was to create an outlet for creative music. We were creating our own work and we did quite a bit of work, actually, during those years, up to a point where we worked six nights a week. We got our own theater and put on some really wonderful musical programs. But that was the whole idea, that we weren't being heard, we weren't working, and no one was hiring us. So we felt we had to create our own type of situation. And that's what we did."

One of the many groups the AACM spawned was the Art Ensemble of Chicago, which was first known as the Roscoe Mitchell Art Ensemble and was renamed upon the group's expatriation to Europe in 1969. The original members, in addition to Mitchell and Lester, were reed player Joseph Jarman and bassist Malachi Favors. Drummer Don Moye joined the group after they arrived in France.

"We had played all over the U.S. and Canada and we had made three or four records before going to Europe," Lester recounted. "We wanted to play what we wanted to play, we wanted to devote our full time to this, and the only way for us to do that, we thought, was to go to another country where the music was more readily accepted. You see, at that time I was with the Art Ensemble one day and Jackie Wilson the next. We just packed up everything that we had in the States, just moved everything, families, kids, cars, motorcycles, everything, took it all to Paris."

Although none had been to Europe before as a musician (several had spent tours of duty abroad in the armed services), the combo was working every night before the first week was over. "We leased us a big house in the country," Lester reminisced, "and word got around immediately. Like I say, we had recorded and our reputation preceded us. We landed this gig in a theater and that was our base for a long time. We traveled all over Europe for the governments over there, we did quite a few concerts all over, festivals, workshops, everything. I mean, it was quite different from the reception we had gotten in the United States.

"We were on the road all the time, always a traveling group, three trucks and two motorcycles. We left France after one year and were on the road *literally* for three or four months," said Lester, describing a routine of cookouts along the side of the highway and tenting down at night. "We were *run out* of France, actually. While we were there the whole revolutionary deal was going on here and there was an interview on the radio about us. Not with our participation, just about us, and they were saying, 'Yes, this is the new black power and this great revolutionary band happens to be living here.'

"Anyway, whoever the count or the earl was of the area we were living in just happened to be listening and he said, 'What're they doing? The Panthers! They've moved into my area!' The next day the police were at our door and said, if we didn't leave, they would escort us to the frontier. So we said, 'Well, we just happen to be leaving anyway.' "

For all of the open-arms welcome the group was accorded upon their arrival in Europe, Lester said that he did not feel that either their audiences or the critics comprehended what the group was all about. "It's much better now after almost twenty years, but at first it was like they didn't know what we were doing. They'd say, 'Lester, we don't know what he's doing, he's just doing something different. But it must mean black power.' "

The Art Ensemble of Chicago has since those early years abroad become such a regular visitor to Europe that a performance there draws thousands. "I think we draw our largest, wildest audiences in Italy," said Lester, "the further south you get. As far as the acceptance of the music, it's pretty much the same from Spain to Norway. They all seem to like it. Forty-five minutes or an hour after we finish a concert half the audience is still there discussing what we did."

However, Lester hastened to point out that the high regard that Europeans hold for jazz "is not because Europeans are any more cultured, any more intellectually advanced than Americans. Americans have more understanding when they hear jazz than anyone else, and I think some of our best audiences are American audiences because they can relate to the music. They just don't get a chance to hear it here.

"Over there they don't have this large black population to deal with, so they can deal with the music intellectually. They don't want to deal with it here, so they try to demean it, belittle it. Here they don't want to hear the music of 'these people who haven't made any contribution at all.' In other countries I tell them, 'Put forty million blacks here and see how much jazz you like.' "

Having observed the Art Ensemble in action on a number of occasions I was curious what the ingredients of a performance were, what part the tradition played in the group's ethos, and how Lester himself conceived of his own role in the ensemble.

"We've got everything that has traditionally been used," he began. "We rehearse and we have cues, we have arrangements, we have extensions of arrangements, we also move on, we segue into different things. It depends on what the composition is. If we have a long composition, we try to convey a certain thing, we think of *that*. It's a lot improvised, it's a lot also written, and I'll defy anyone to tell me which is which, because some of the parts that people think are improvised are written and some of the improvised parts are head arrangements. It's a matter of putting it all together, of group consciousness. We try to not be limited to just playing in a form like ABA, take a couple solos and that's the end of it. We're about opening up because we feel that the music has been so limited, everyone is so closed up, so afraid to do anything.

"We are trying to do an extension, trying to take what has gone before, mill it around in our minds, add some of *us* to it, and then— this is *our* version of what has happened. When I play my horn, I'm not satisfied to repeat or copy Miles or Diz. I don't believe that that's the test of the truth of the spiritual part of this music. I believe that you have to contribute after you've learned *your* lessons. What *I* play is *my* opinion in regard to what has gone on before, because without that, we wouldn't be here. Diz *invented* [his] way of playing. He listened to Louie and he listened to Roy and then he came up with *his* contribution. And I've tried to continue that situation. That's the way anything is done, not just music."

As for the truckload of instruments that the Art Ensemble hauls around and clutters up a stage with, Lester explained that the group "tries to explore different colors. When we first started coming into being, one of our main focuses was to be more expressive, to create a music that was more expressive than what had been played in the past. It involves having other colors, other dimensions to colors, other textures. I guess the difference is just like for a painter to have three colors available or having a hundred colors available. He can do much more with the hundred different colors."

One of the new colors recently added to the ensemble was that provided by their new synthesizer, which Joseph Jarman was handling. "We used to have a theremin,"* Lester pointed out, "and you waved your hands in front of it. It was Joseph's also, but it got stolen in Brussels. We use a synthesizer like we use another bell. We're not running everything through it. We can do some things with the synthesizer and some gongs and shit on top of it, man, and it's unbelievable!

"I'm going to play trumpet until I'm sixty," Lester assured me, "and on my sixtieth birthday I'm retiring. I'm going to do limited touring up until then, because I've learned that I can live exactly the way I want to live and I really don't want to walk on that stage every night playing for a living. I'll probably be involved with music and playing somewhat for the rest of my life, in a teaching situation, maybe lecturing, or just writing books or something, but I'm not going to be walking on that stage every night.

"I'm much more concerned about other musicians than I am about myself," Lester declared when I inquired into his assessment of the current state of the art form and its practitioners. "It's just that I hate to see someone like Kenny Dorham waste away for years and then die like that and then no one recognizes him, or if they do recognize him, it's long after he's gone. Because, you know, the music could be so much better. I mean, what if Kenny Dorham had been kept healthy, what would he be doing now? What could Bird have done, if he could have stayed healthy? Where could the music actually be now?

"Our music is a kind of music that's good music *in spite of* the hardships, *in spite of* the fact that you can't go to the dentist to get your teeth fixed, *in spite of* you can't afford a decent instrument, *in spite of*

---

*An electronic instrument consisting of oscillators, rod, loop antennae, and a speaker, named after its Russian inventor Leo Theremin.

the racist conditions. It's still a great music. But how much really *greater* could it be, let's say, if we were to receive the funds of a symphony? You could add all the black musicians up, I mean all the jazz guys, total all of them up, and then we don't get as much support as *one* symphony orchestra, the New York Philharmonic. But at the same time, we still play a vital music and a true American music.

"I've been in this business all these years," Lester lamented, "and if I can make as much money as my mailman, that's a good year for me. And *I'm* well known!" Indeed, at the time of that observation Lester had been a professional musician for nearly three decades and had been playing the trumpet for close to forty years. "One thing they always tell the young students, see, is that they have to be ready to die, if they're going to get into this."

Citing pianist Albert Dailey, who had recently died at the age of forty-six, Lester pointed out, "Society has said that jazz musicians aren't really worthy, they're not really 'musicians,' they're a bunch of hop heads. The problem—let's see, how can I explain this?—is about the amount of talent and the degree of acceptance. Albert Dailey could have been the head of Juilliard, you know what I'm saying? But instead he was playing in taverns night to night. You have a musician of that high quality, that talented, and no one knows about him. Most musicians—I'm talking about the jazz musicians, not guys who work in the pit band—they don't have money to have things like normal people. We don't get social security, we can't get unemployment compensation, and the hospital doesn't care about you if you don't have money or insurance. It's really tough."

Nevertheless, Lester insists that, in his own case, the hardships have given him strength, have helped mold him as an artist. "Playing jazz is more than technique. It's about expression of feeling and it's about memories, it's about a lot of things. You're trying to express emotion and you have to have a lot of feeling to be able to convey this. I was forced to have a variety of experiences and all of these things helped me as far as trying to put myself into the music."

Yet another artist whose approach incorporates a wide spectrum of the jazz idiom from the blues to free playing is the El Reno, Oklahoma–born multi-reed and flute player, pianist, composer, leader, and educator Sam Rivers, who grew up in Little Rock.

"My father and mother are both musicians and my grandfather and his sister were musicians," Sam Rivers told me, referring to his pianist mother, to his father's membership in the Fisk Jubilee Singers and Silvertone Quartet, and to his grandfather Marshall W. Taylor, a distinguished minister who in 1883 published a hymnal titled *A Collection of Revival Hymns and Plantation Melodies*. "Three generations of black American musicians. So I consider myself fairly well immersed in spirituals, the swing era, bebop, and avant-garde.

"My music is traditional, yet it's also avant-garde and it swings with a feeling, has blues, then it's also 'out' and screaming. I really try to run the gamut of emotions and this is what music is all about. The feeling is the most important thing," Sam observed, additionally citing the sentiment expressed in the title of Ellington's "It Don't Mean a Thing If It Ain't Got That Swing" as still another basic truth, the need for that mysterious element of "swing," an element that truly defies analysis. "I learned years ago, no matter how technical a guy was up there, if one guy comes up behind him and holds one note and brings the house down—I mean, the feeling is what is supposed to be projected.

"I consider myself also a musician's musician, reaching out to project emotion to the public without playing down to them, trying to bring them up a little, trying to give them a little of something that they've heard, but a lot different, and also something they can pat their feet to. If you're creating spontaneously on the spot, you're definitely going to make some mistakes. In other words, it should be like uncharted territory, like every time you take a photo. We're not playing what we rehearsed, we're out there playing something that, if you heard it yesterday, we try not to play that tomorrow. It's the direct opposite of the symphony player. It's a different kind of approach. We both make concert music, only one is a European derivative, the other is American.

"The jazz musician is trying to be personal, trying to sound as an individual. On the other hand, the symphony player is trying to be *im*-personal, he wants to be a part of something, not to stick out. The jazz musician *must* stick out. His whole thing is projecting his particular idea of music, that it is to be played in order for a person to *feel* it. If jazz musicians *don't* make mistakes, then I'm not sure whether they're attempting anything creative or new at that point. But as far as what's written—no, they don't make any more mistakes, I think they're just as exactive as classical musicians. The jazz musician is more noted for his improvisations and his creativity than he is for the written. That's what makes it jazz."

Sam attended the Boston Conservatory of Music in the late 1940s and early '50s and subsequently became a fixture on that city's jazz scene, working throughout the '50s with Jaki Byard, trumpeters Herb Pomeroy and Joe Gordon, saxophonist Gigi Gryce, and others. Sam also toured with Billie Holiday and bluesman T-Bone Walker. In 1964, the year he settled in New York, he became a member of the Miles Davis Quintet for six months, visiting Japan with the group.

"In the '60s I was pretty much in between," Sam said, attempting to pinpoint where he was, stylistically, upon joining Davis. "I'd been introduced already to the avant-garde of Cecil Taylor and Ornette Coleman and Andrew Hill and other musicians that were working in the avant-garde, like looking into that, but I was also playing the standards and had been playing bebop professionally for, say, ten years or so. I was trying to put these two things together, bebop and avant-garde, rather than be completely bebop or completely avant-garde. So I tried to find a nice middle ground between the two and it's pretty much worked successfully for me. It was at the time that I joined the Miles Davis Quintet, after George Coleman quit, and I did a world tour with him."

In 1971 Sam Rivers and his wife Bea founded Studio Rivbea in a loft in lower Manhattan, and the studio became a veritable university of contemporary jazz studies for the rest of that decade and for several years into the '80s when, because of a heavy travel schedule and other commitments, Sam was compelled to give it up.

"I'm just being realistic in saying that most musicians that are on the international scene today are there in part because of their work at the studio, when I had it in the bleak '70s when no one was listening, no one could get a record date or anything, and there was no place to play. At Studio Rivbea they were able to be heard by some enterprising record people and entrepreneurs and so now they are international. It was just that I happened along at the right time and was able to make a contribution in that way. And from that we have the fruit of the musicians that are on the scene as the contemporary musicians of today, because of the fact that I was fortunate enough to get the studio set up in the '70s with some help from the government.

"Everything is pretty structured in the U.S.," Sam pointed out. "It's very tight. They jump on what's making the money and everything else is excluded. So the music at all levels suffers because of that. What we play now is pretty much high art form so it falls back on the government to preserve it in the same way that we preserve our symphony orchestras. Now we ought to start preserving our jazz artists. It's pretty much a duty of the state to do so."

Sam Rivers was still maintaining a busy schedule of performance, composing, and teaching in the late 1980s. Among the institutions where he has instructed are Connecticut College, Wesleyan University, Antioch College, and Seattle's Cornish Institute. The formats he has employed over recent years include solo performance on saxophone, flute, and piano, duos, quartets, and other small combos, an aggregation of eleven saxophones, a woodwind ensemble, a fifteen-member big band of conventional make-up, and symphony orchestras with which he has been featured as guest soloist. In confirmation of his across-the-spectrum abilities, Sam joined some sixty other, mostly mainstream, musicians for a 1988 "jazz cruise" aboard the *Norway*, during the course of which he joined Buddy Tate, Arnett Cobb, Illinois Jacquet, Flip Phillips, Benny Carter, and Norwegian Totti Bergh for a wall-to-wall saxophone jam.

"It's still wide open," Sam concluded our discussion. "It's still available, still completely free, to come up with some new style of music at any time. It comes from the imagination of man."

The trumpeter Malachi Thompson fleshed out the account provided by Sam Rivers. I asked Malachi to describe the so-called "loft" scene in New York in the 1970s.

"When I first moved to New York back in 1974," Kentucky-born and Chicago-raised Malachi commenced, "one of the first places that I lived in was a loft over on the East Side. Matter of fact, there were musicians living on each floor. There were three floors and a business on the first floor. There was a percussionist who used to play with Miles Davis on the ground floor and on the second floor was a saxophone player and me and drummer John Betsch were sharing a loft on the third floor. Out of this we started a band and it was called the Tenth Street House Band. We started doing loft parties. We'd invite our friends over and we would play.

"At the same time there were other lofts that had been around for like longer, like the Ladies's Fort, which was Joe Lee Wilson's place, one of the more popular lofts among musicians—I did one of my first gigs in New York there—and Sam Rivers had a place that was right down the street. Further downtown Ornette Coleman still had his loft and later on Rashied Ali renovated a loft and changed it into a club called Ali's Alley. That was probably the biggest loft and the best-equipped place. Oh, yeah, I can't forget Studio We has been around the longest. They started back in the mid-'60s. It might have been around for close to twenty years now and they're still happening. It has a nonprofit organization that keeps the thing going. I think Studio We owns the whole building and that's one of the ways it was able to survive. Sam Rivers had nonprofit status. Others went that route because it was part of their whole survival mechanism. So these were the main lofts and there were the lofts on Broadway called the Environs that was doing stuff on a regular basis.

"These lofts were like old factory buildings, warehouses and stuff, and all these businesses might have been moving to Jersey or something and there was a lot of empty space. So what happened was rents were cheap and musicians would take advantage of that for their studio and space. Depending on the space, rents were anywhere from two hundred and fifty dollars on up to, I guess, a thousand dollars. A lot of those places weren't really designed for living, and most of the work that was done, as far as renovation, was done by the musicians themselves. You know, guys would chip in and help to fix the place up.

"That whole loft scene, what it was all about, was a lot of the younger musicians weren't able to get into the clubs, and at that time there wasn't really a whole bunch of clubs happening. And so the loft scene was a kind of alternative place for the musicians to write their music and get it played and other musicians to get it together and expose themselves to a wider audience without going to the clubs. 'Cause mainly at that time clubs were only dealing with established musicians who, say, were out of the '60s and the '50s and before that. The Village Gate and the Village Vanguard weren't too interested in that second generation of cats coming through.

"The main purpose, like I said, was for the musicians to get together and showcase their music out of the club atmosphere. What happened was that the public could go because it wasn't as expensive as going to a club and there wasn't any waiter or waitresses hassling you to buy drinks. It was a really relaxed environment, the musicians were accessible, you could sit around, and people would sit down and talk with the cats. Usually there wasn't any liquor served. They would have fruit juice or sometimes there would be wine, sell wine out of a jug or something like that or give the wine away as part of the admission fee.

"A lot of times, after a gig, the cats would get together and go to a loft and hang out and jam all night. It wasn't so much a private party as it was a self-help kind of program, educational in a sense, where guys could get together and play and experiment with their stuff without being under the pressure of being at a club or having to do so many sets a night and having to worry about attendance and drawing and all that stuff. It was like a less pressured situation.

"They would advertise in the local press, mainly in the *Village Voice,* and once they really got started the press jumped on it and they coined the phrase 'loft jazz.' Actually, there was really no kind of music called loft jazz that was only played in lofts, there was just music that was played in lofts, all kinds of music, but the music did tend to lend itself to the newer stuff, the more inventive composers.

"All of a sudden, it became a whole scene and people started coming from Europe and from all over the world, hanging out at the lofts because they found that that's where the real new music was coming out of, because that's where the musicians were. After a while it became like a very fashionable type thing and the rents started going up and it became more a middle-class type situation with the lofts. The fact is that the musicians made it popular, along with the press. The press has to be given credit because they did give it a lot of support and a lot of the loft events were covered by the *New York Times.* There were some recordings made at the Ladies' Fort and there was a series of recordings made at Sam Rivers' loft called 'Wildflowers.'

"Out of the loft scene of the '70s a lot of musicians built their reputations, people like Arthur Blythe, David Murray, and myself. Towards the late '70s Sam Rivers lost his place—that was one of the first places to go—and the Ladies' Fort changed hands a couple of times and then it kind of faded out. A lot of the action shifted over to Rashied Ali's place. It had become a club and actually over a period of time he had a very popular jazz club, Ali's Alley, well attended, and it was like a six-night-a-week club."

Malachi Thompson became disillusioned with the New York scene in the early 1980s and moved to Washington, D.C. "In New York you get into a thing of being non-emotional," he opined. "I've seen people there step over people lying in the street to get to their Rolls Royces." He added, in explanation of his preference for the slower pace in the nation's capital, "I was born in the South so this is my kind of groove."

Gary Peacock, who grew up in Idaho and was living in Seattle in the 1980s, is considered one of the great masters of the bass in jazz and is a recognized innovator in terms of the expanded role his instrument began to assume in the 1960s, a role which is taken for granted today as a legitimate approach in contemporary jazz expression. Gary's associations from the 1950s have included a diversity of musicians, including Barney Kessel, Paul Bley, reed player Jimmy Giuffre, trombonist Albert Mangelsdorff, pianist Bill Evans, Miles Davis, Don Cherry, Keith Jarrett, saxophonists Sadao Watanabe and Albert Ayler, and Don Pullen.

"I can remember when I was in Los Angeles about 1959 and I noticed that I was playing less time," recalled Peacock. "In fact, playing the time became very irritating, it just felt like a straightjacket and I wanted to kind of break through that. I simply started listening in myself to what a bass could do and I just began to intentionally play out of time, with some notions, some glimmers of what could be done. I noticed that the time could literally be there and I could be playing something that was basically ametric and yet it worked in some way as far as my ears were concerned.

"As far as a lot of musicians I was working with, it didn't work at all for them. They'd say, 'Gee, what happened to you? You used to be able to play so well, now you don't know what the hell you're doing!' It was just a point where I couldn't in all honesty to myself continue playing time. So it was a little bit rough and I finally moved to New York. One of the aspects of moving there was that I knew that there would be a wellspring of musicians to draw upon, to expand further and develop more."

Gary expatiated upon just what constituted his expansion and development. "A lot of it had to do with the shift in the context for improvisation. Prior to [the 1960s] the role of the bass was primarily a support role. So if a bassist could handle time and play changes and have some kind of sense of swing, then the bass player would recognize this and the people around him would recognize this as the purpose of the instrument and that's what it was all about.

"An interesting question [came up]: What would happen if we didn't play the time but just allowed it to be there. Instead of going, 'ding-ding-ding,' just allow it to be there and play something. So that began to, in some ways, be experimented with by a lot of musicians in the late '50s and early '60s. It became obvious that the bass didn't need to play 'ding-ding-ding' on every beat. But also, if the bass didn't do that, it created a different context for those who were playing around him in a group. It opened up something for the pianist, it opened up something for the drums. It had some kind of interchange. Obviously, when the bass was playing in time and the drums in time it became very mechanical. So it began to open up the role for the drums and the piano. At the same time, the bass would be handling the harmonic aspect, playing the roots of the chords. But instead of playing them on beat every time or in some kind of metric form, it was more flexible."

Playing a gig one night in Philadelphia in the mid-1960s, violinist John Blake heard about "a teacher who played the violin between his legs." John sought out the teacher, who turned out to be an Australian who had lived in India for a decade, and for five years John became his pupil. Thus did John bring a new voice to jazz violin, combining his early classical training, the Philly r&b sound, and the microtonal approach of the South Indian Carnatic violin school.

"It has the effect of making the instrument being able to cry," John explained, "and to get very close to the human voice. In playing blues or anything that has a crying kind of feeling to it, I find that that technique has given me the control to be able to express that at any moment that I desire. Take blues singers—they never sing directly on pitch, they bend the notes. That bending, that sliding, is very much in the Indian tradition. So when I studied Indian music and then come back to playing blues, it took me a while to learn how to slide in a variety of ways so that I could have that kind of control. So when people hear me play, they're not really hearing Indian music, what they're hearing is a blending of cultures. I've been able to integrate some of those techniques into my style. It helps me to be able to express what I'm saying much, much better. There's nothing else that's helped me as much to have a sound that's identifiable, so that someone can say, 'That's John Blake.' "

Another who went far afield for fresh inspiration was former saxophonist Rufus Harley, who like John Blake had a Philadelphia background. "I was watching the burial services of President Kennedy on television," Rufus told me, "and it was the Black Watch Bagpipe Marching Band of Canada that inspired me. Their sound sort of stuck with me and I bought the set of bagpipes that I have out of a pawn shop for a hundred and twenty dollars. Then I sort of investigated the instrument and brought it up to date to modern music. I had to get some idea of traditional pipe playing but I found that if I played it like I would play an open-hole flute or an oboe, or a clarinet, I would get a better sound and the pitch would come truer. I also reversed the positions of the hands.

"I say that the bagpipes go way back to the beginning. It's the mother instrument. Everybody is trying to understand the elements of nature and the bagpipes are designed right after the anatomy functions, the human heart."

Having taken up the bagpipes, Rufus became the very first jazz bagpipe player and he went on to record on the instrument with Herbie Mann, Sonny Stitt, Sonny Rollins, and his own groups. Were we to continue on the theme of exotic sources and nontraditional, even non-Western instruments being introduced into the jazz mainstream, we would be compelled to credit the contributions along these lines of Don Cherry, the quartet Oregon, John Coltrane, and others. Suffice it to say that the musics of Africa, the Middle East, India, South America, Native Americans, and European folk cultures have all made an impact upon the art form of jazz these past four decades or so to a degree that dwarfs their earlier presence in the idiom. Of exotic influences, perhaps only Jelly Roll Morton's "Spanish tinge" can lay claim to being a significant element in jazz before, say, the 1940s.

Speaking of Jelly Roll Morton, the tradition remains, as we have often remarked, the lode which the jazz artist must continue to mine. It has in the 1980s become almost commonplace to discover that a new recording by a contemporary, so-called avant-garde, group includes a selection from the classic jazz repertoire of the 1920s or '30s, but the trio Air, which came out of the AACM in Chicago in the 1960s, was as early as the mid-'70s performing its own free versions of compositions by Jelly Roll Morton, Scott Joplin, James P. Johnson, Fats Waller, and Duke Ellington.

Henry Threadgill, multi-reed and flute player, composer, and member of Air (along with bassist Fred Hopkins and drummer Steve McCall, who died in 1989 at the age of 55), pointed out: "I don't particularly like the use of the word 'jazz' because it means so many different things to different people. I reject most of these terms simply because they categorically put you in some situations where people don't listen to you because of a tag that's on you. These labels create isolation. So I say the best term is just 'music'—and contemporary music is avant-garde. It's not laboratory music—what we're creating is for people as the receivers of it."

Or take pianist Don Pullen, who in the late 1980s was beginning to be acknowledged as one of the several chief contemporary creators in jazz on his instrument. "I had an organ group for years. During the lean times that's what really kept me going. At that time just about every little bar had an organ in it so I was able to survive doing that. I'm the same person, basically, because I still incorporate all those influences in what I do today, so that my music has become more varied and more interesting for me. Each time is like a new time and I look forward to playing.

"I don't really play any differently than I did before, except in terms of growth and understanding. It's simply that I know how to use the same tools and techniques in a more natural way. The artist has a responsibility to bring his music to the people in a way that they can understand it, at the same time not losing any of its high ideals and quality.

"People don't really think that I'm doing anything that's so different," Don pointedly informed me, "but as far as writers and critics are concerned, I think they have a little problem with it. Periodically, someone will write something about me that seems to be so dated, really out of tune. But I think the writers are becoming more educated as to what I do, beginning to dissassociate me from their past assumptions. Events are forcing them to review their opinion of me.

"One thing that makes the music more acceptable now . . . is that the level of musicianship has improved. . . . Avant-garde opened the door for people to enter the music world who otherwise could not have been in it, because of the freedom that it offered. As time went on, the door closed in front and behind them and they really didn't have anywhere to go because their musicianship, whatever there was, ran out. A lot of those people were left by the wayside and those who survived are the real musicians from that period who are able to keep the music moving and growing and relating to people. A person should play from his own experience—that's the only thing your music can talk about."

Drummer Andrew Cyrille, who was born in Brooklyn, New York, talked about the several traditions he draws upon. "The thing that impressed me most about Baby Dodds, Chick Webb, Sonny Greer, Gene Krupa, Louie Bellson, and Roy Haynes is how they go about making a drum solo a viable piece of music with a beginning, a body, and a conclusion.

"I look for the same thing in a drum solo that I look for in a violin solo or a piano solo or saxophone solo. I just did one for an hour over in France two Saturdays ago. I get some motifs, I have some idea of where I'm going. I'll make a chart for myself, whether I'll memorize it or have it written out, and that will suggest to me at some points that I do something else. I'll say, here I'm going to play this kind of rhythm, the next time I'm going to play that type of rhythm. I'm either going to play fast or I'm going to play slower, I'm going to play softer or louder. I might play with a lot of space. I might choose one area of the instrument to play, like for instance the cymbals, or I might decide to play on the tom toms, or the bass drum. Also, I will do certain things to the drum head in order to vary the sound. I've heard Joe Morello play 'Mary Had a Little Lamb' on one of his tom toms.

"So there are a number of things that you can do in terms of elaboration. It just depends on what you decide to do and how you go about it, how you evolve it. I try to make it as viable a musical conversational piece as I possibly can, like I'm talking to you now.

"From the African matrix, there's a lot to be said about structure and how they improvise on structures and how they put drum choirs together, how they will have two or three drummers support a lead drummer, how they think about drumming in relationship to dance or to certain religious practices, how they are trying to invoke spirits to do certain things, or at least to hear them with their pleas. Musically, what happens is they go ahead and construct and work together. It's the same thing that happens when I work with a drum choir.

"In terms of African rhythms, there are lot of things that can be said. It can be polyrhythmic, polytonal, there's a large degree of improvisation, room for individual solo, a theme with a beginning, middle, and end, perhaps a statement. Like in English, we might have a main clause and a subordinate clause, the antiphonal kind of thing with questions and answers. A lot of these things I use not only in my soloing but also when I do drum duets and trios and quartets. It's just a matter of how you parcel those ingredients around."

The trombonist Roswell Rudd serves as one of the better illustrations of the inseparable relationship between the old and the new forms of jazz. Roswell's entry into jazz performance was as a member of a college Dixieland group, Eli's Chosen Six, at Yale in the 1950s. After that, and into the '60s, Roswell went on to work in traditional and Swing Era style bands alongside such stalwarts of those idioms as Wild Bill Davison, Eddie Condon, and Buck Clayton. In the early '60s he was for a while concentrating in one combo—along with fellow graduate of the traditional school soprano saxophonist Steve Lacy—on the music of Thelonious Monk. Around the same time he worked with the bebop pianist Herbie Nichols for a couple of years. Then, in 1962, he began a series of associations with musicians of the avant-garde of that time, compiling a roster that eventually included the likes of Archie Shepp, Milford Graves, Bill Dixon, Charlie Haden, Beaver Harris, Karl Berger, Lee Konitz, Robin Kenyatta, Lonnie Liston Smith, John Tchicai, Dennis Charles, and Gato Barbieri.

"What we call the 'avant-garde,'" Roswell explained, "starting around the end of the 1950s and on up through the '60s, in many ways was a return to the past, a return to big melodic statements and emphasis on different kinds of rhythm and on—and this is where the Dixieland helped me—collective improvisation. In bop the emphasis had been primarily upon ensemble performance of composed stuff, and improvisation was done with a rhythm section and usually one horn player. What people called avant-garde music in 1960, for the young players, may in fact have been some kind of extension of Dixieland playing or New Orleans playing. If you take it further back than that, it goes into folk culture and back into some ancient, ancient times.

"I grew up on the recordings of Louis Armstrong, Jelly Roll Morton, Duke Ellington, and Benny Goodman," Roswell pointed out. "And then later I went into a sort of music study thing and began to compose and research a lot of things. I wouldn't exchange those times for all the tea in China. That was where I got my real education," he insisted, in reference to an extended period beginning in the mid-1960s during which he was deeply involved in ethnomusicological research and was a professor of that subject.

"It wasn't as if I had one foot in one world and one in the other," said Roswell, clarifying his crossover role in jazz. "The old music is still what I do. I haven't played recently in any one hundred percent Dixieland combos, but some of the individuals I play with now have that kind of background and at times the music can sound that way. For instance, with this quartet"—Roswell was in town on the occasion of a gig with keyboard player Terry Adams, bassist Joey Spaminato, and drummer Tommy Ardolino, who constitute NRBQ, an r&b trio—"we play Herbie Nichols' '52nd Street Rag' and we really *play* it, we really do! We play a lot of the old music and we're adding things. It's creative more than *re*-creative. It's what I've always been doing and always wanted to do and now here are three other guys that can do it."

Roswell was still playing and composing in 1990. It must give him no little satisfaction that a number of prominent young trombonists of contemporary jazz—for example, Steve Turre, Ray Anderson, Frank Lacy, Craig Harris—have chosen to incorporate into their instrumental voices the earthy "gutbucket" attack of the traditionalist, which is Roswell's signature sound. Incidentally, when I last spoke with Roswell he told me that he had taken up the study of classical Greek. Perhaps the rhythms of that tongue's sublime poetry will one day turn up in Roswell's blowing!

For a fitting coda to this chapter on post-bebop "avant-garde" developments in jazz, a development that has been as replete with references to the past as any of the several other major styles or periods of the music, we turn again to Arthur Blythe for yet another soulful "change" on his favorite metaphor.

"You always have to maintain a little cornmeal in cornbread, regardless of whether you put carrots in it or nuts or prunes or raisins. And I think that it's also essential that you have the tradition in all that you do."

8

Singers

You have to live the blues and I think I've
lived a little bit of it.

Carrie Smith

If I go too long without performing I get
crazy.

Shirley Horn

I don't get up there and try to show you how
much I'm waiting for the man I love. I get
up there and try to wait for the man I love,
and just let you in on it.

Barbara Lea

No general discussion of the jazz vocal art would be complete without inclusion of the role that the blues, specifically the vocal blues, has played in the jazz idiom. To remark that that role was seminal is an understatement. But before we move on to hear of the formative experiences of several blues singers, let us first consider the relationship between vocal and instrumental expression in jazz, for virtually all blues singers and a great many jazz singers have utilized elements of instrumental attack and have simulated something of the sound quality of horns. We again call upon jazz historian William J. Schafer.

"I think this is another strong point that runs all the way through jazz history," Schafer opined in a conversation with me, "and that's the connection between vocal music and instrumental music in a really close interrelationship. It really comes out, I think, as much in the [New Orleans] brass band tradition as anywhere else because you see the connection between the 'read-as-written' parts of the music, the head or ear arrangements that the bands used, and a sort of common source that all the musicians had, whether they were trained to read music or not, in the church music of the time.

"Bunk Johnson said one time that he could play anything on his horn that he could whistle and I think Louis Armstrong made a very similar sort of remark, that if he could hum something, he could play it on his trumpet. And you get that sort of connection, I think, very vividly through this fund of church music, especially the hymns and spirituals that come out in the funeral repertory of the brass bands. And it comes out in the sound of someone like Louis Armstrong, who I think is the greatest example of this. The connection between his singing style and his trumpet playing is just absolute, identical. I think you hear this through the instrumentation of the brass band, too.

"It's a very close relationship, and one of the things that really distinguishes jazz from a lot of other forms of vernacular music in this country is that the two things have been almost an identity, not one an accompaniment to the other or one an opposite version of the other."

The themes introduced in those foregoing paragraphs will be repeated, specifically or by implication, throughout this chapter in some of the accounts we shall draw upon. Let us hear, first of all, from several blues singers.

"I was surrounded by music," Tennessee native Johnny Shines told me. "My mother had a brother which his whole family was sanctified. Everybody in his family played music and my mother she also played the guitar. She played quite well but she was more of a strummer than a picker, see. I was surrounded by church music. Matter of fact, when I first started to play, and I started playing organ, everything I played was church airs, because I was around my mother and that was all she allowed, she didn't allow any blues in the house. My brother, he played, but he kept his gui-tar in the crib somewhere, he didn't bring it in the house.

"Tell you the truth about it, what we call the blues, in the beginning was really what all of your American music come out of, see. There is church music, as we call it, spiritual, jubilee, and on and on, and all of this come out of the bowels of slaves. We were taught a lot of Dr. Watts' songs, which was wrote by a white man, but this was to entice and encourage the black people away from the blues because the blues was a descriptive thing. The blues was taughten by life itself. You take when the slaves was brought over here, they were not allowed to play their instruments or anything like that and when they began to sing, moan, and pray—understand what I mean?—and chatter, which they called 'chants,' and groan to themselves, these are the things that later on was called blues. But how in the world could a person be anything else *but* blue, as we describe the blues of today, how could he be anything else but blue when he was in slavery? The only thing that he had to live with was blue.

"I started playing the blues when I was about sixteen years old," said Johnny, resuming the story of his early life, "really got into the blues thing. I got married when I was sixteen and my parents signed for me to get married. My oldest brother was living with me, . . . a gui-tar player, and he played this song 'Jim String'. I liked it very, very much and we really got to the place where the people would get off on [me playing] 'Jim String' and they called me Little Jim String, see, for a long time.

"Then I sit down one night, I was lookin' at Howlin' Wolf play. He's a great inspiration to me, he was a mountain of a man in my eyesight, and I thought there was nothing no greater than Howlin' Wolf. And, as you know, when he passed there was none no greater as a blues-man in millions of people's eyesight. And sittin' there lookin' at him play, I say, 'What *he's* doing, *I* can do it, too,' because I knew his tunings, you know. But I didn't know his riffs, I didn't know his licks, I didn't know anything, see, but sittin' there lookin' at him, I visual-ized everything he was doing.

"So Howlin' Wolf, he had a habit of sittin' his gui-tar down, which he stopped from that night on. He set his gui-tar down and he went to the crap table or whatsoever, and when he come back I had the house jumpin', playing' *his* songs. I was playin' and singin' his songs. So then they called me Little Wolf for a helluva long time.

"Later I played in Hughes, Arkansas, which was the same place that I met Wolf, and I was playin' with a boy and he kept crowdin' me about the idea of meetin' this fellow named Robert Johnson. At that time I could sing like a bird, but I didn't know there was another bird, see. At that time we had a little game going we called 'head cuttin'.' We looked for better guys to compete with them, see if we could cut their heads, see, which was, 'I'll play them, gain better applauses and better influence among the public' than they did. Then we would achieve the idea that we cut their heads, see, and so did the public. We called it 'head cuttin'.'

"So I finally made up my mind I would meet this fellow Robert Johnson. He was a great recording artist and everything. So I went to Helena, Arkansas, where he was. I met him, sure enough. It was a *helluva* jam session, man," Johnny recalled, still in awe of Johnson's presence at that mid-1930s meeting. "It was something terrible! At that time I had learned to play in four or five different keys and I was studying very hard, nothin' but the blues. And I met Robert, but he was too much for me. He was too much for me. So then I wanted to learn what he knew and I followed Robert just about all over the United States, all the way up in Canada, and back into the United States and back to Helena, Arkansas. But he went to Mississippi and 1938 was the last time I saw him."

The circumstances of Johnny Shines's introduction to the blues can serve as a paradigm. Still, we choose to briefly supplement Johnny's story with those of several other blues artists before we make a transition to jazz singers.

"The guys who worked the farms would moonlight," Bill Harris told me in one of a number of interviews. He was talking about his 1930s childhood in Nashville, North Carolina. "They'd bring their guitars into town, get on the corner, and play the blues. They would come there to make some money, especially around the time when the farmers auctioned off the tobacco. When I'd go to the store on Saturday I'd see them and it fascinated me the way they played. They had names like Laughing Lanky and Billeye and they never recorded, but they could real-ly play the blues."

One Saturday it was too much for the eight-year-old blues-struck youngster. Arriving home, he ripped some wire off the screen door, tacked it to a cigar box, strung it out on a broom handle, and tried his own hands at the blues. Before another year was up, an uncle sent him a guitar. Because the neck of the instrument was warped, Bill was compelled to fret it with a knife blade. "So I learned bottleneck technique without even knowing what it was," he said, chuckling.

For several years Bill had already been picking out tunes on the organ in the church his family attended, and the small repertoire of spirituals he taught himself made him a regular performer at Sunday services there. He also played a homemade set of drums in his high school band, sang, and did a little acting.

"The blues was always around me, but I wasn't supposed to play them. The uppity people, they looked down upon them because they were plain, lowdown, dirty blues. In my family's case, they were negative for religious reasons. You see, I wasn't supposed to get the blues because my daddy was a sanctified preacher. Every time he heard me sing one of those Bessie Smith things my brother'd scream on me, 'Mama, Junior out here singin' 'nother one o' them reels!'

"Yeah, my daddy made my guitar teacher Mr. Blake promise to teach me only the spirituals. Mr. Blake looked at me one day, said, 'Boy, you want to play the blues, doncha?' I said, 'Yessuh!' He said, 'Well, you gotta go like this.'" At this juncture Bill picked up his guitar to demonstrate Mr. Blake's pedagogical deviation from Preacher Harris's stricture and elicited from the instrument some very mean blues in self-accompaniment to, "'Oh, man, in the mornin', this early in the morn-n-in, I ain't got me nothin' but the blu-u-u-u-u-u-ues,'" holding onto that final word until it cried out in pain.

"My first real blues teacher was old George Doughtrey, five, eleven-and-a-half blues shouter. He didn't know he was my mentor, but at my wakin' hour three-thirty in the mornin' I could hear old George leavin' his girl friend's house singin' the blues. Yeah, George was hollerin' and screamin' and whistlin' the blues to keep the 'hants' off of him.

"I'd like to give you my version of George doin' his version of Arthur Big Boy Crudup doin' his version of Peetie Wheatstraw doin' his version of Leadbelly's 'Settin' on Top of the Cotton Pickin' World.' " Upon the conclusion of which majestic intro Bill was off again, whistling the blues to his guitar and then coming in on voice. A voice, incidentally, that could rumble, keen, yodel, quaver, growl, and even croon the blues with a fluency that only a mere surviving few authentic blues artists like himself could still do in the 1980s.

> Lawdy, you know, I doan worr-r-ray
> 'Cause I'm settin', I'm settin'
> On top o' the wor-r-r-r-l-ld.

"Gui-tar Bill" Harris, a mainstay on the blues and jazz scene of Washington, D.C., for better than four decades, died in 1988.

"I was born in a little place called Fort Gains, Georgia," Carrie Smith told me, "and I started singing in the Baptist church in Union, New Jersey, where we moved when I was seven. That's where I got my early training from. I was singing old spirituals and my mother and dad had Bessie Smith records and some of Ethel Waters from back in the twenties and I used to listen to them when I was very small. Something sort of clicked and I says, 'Oh, I'd like to do that one day.'

"You have to live the blues and I think I've lived a little bit of it. They say that blues is nothing but a woman that's lost her man or a man that's lost his woman. You can sing the blues anywhere. When everything else fails, you can get over with the blues."

Washington, D.C., bluesman Nap Turner echoes this sentiment. Asked how he felt about an audience laughing at the concluding line ("Remember, I had my fun until my life ran out") of James Oden's "Goin' Down Slow," a fixture in Nap's repertoire, he assured me, "I love it, because it tells me they understand what the blues are all about. I see the blues as music that purges one of bad things and fixes it so that one can go on."

If anyone should know, it would be Nap Turner, who grew up hearing the blues in a coal mining town in West Virginia in the 1930s. An aunt played guitar in Delta style, and Nap said, "The blues was around me all the time." Nap was a bebop bass player until he found himself "in the penitentiary" on a drug bust in the 1950s, the first of a series of incarcerations that continued into the late 1960s.

In prison the blues became Nap's constant companion. "That's when I really started to singin' the blues. I had just lost my family—my first wife had left me and taken my babies with her. They had a little band in jail and the bandleader was a friend I had known for a long time, Little John Anthony—he's been dead quite a while now—and he was a blues singer. He was tellin' me about how music can help to make the time in jail much easier. So that's when I started practicin' singing and doing it.

"It was a metamorphosis, being able to accept the blues as a way that I could express myself and not be ashamed of it. My background was strongly Baptist, fundamentalist Baptist, and although I had been hearing the blues all my life, the blues did not have the status in the community where I was raised as it has now. And learning to sing the blues was a part of that same metamorphosis that happened all across my life, because it was during this period that I really started trying to change my behavior. It took me a lot of years to do it, but I became more in control of my own life, the easier it was for me to accept without any shame the fact that I liked the blues."

Since recovery from heroin addiction around 1970 Nap Turner has held positions in the area of drug rehabilitation and in the 1980s was a narcotics program specialist with the D.C. Department of Human Resources. As a blues singer Nap was still performing in the clubs of his adoptive city in 1990. In 1988 he was one of a number of blues artists from Washington, D.C., who participated in the San Remo (Italy) Blues Festival. At the time of writing Nap could still be heard Wednesday mornings on WPFW-FM (Pacifica) with his radio program "Don't Forget the Blues."

"I started in church at the age of five singing spirituals," said Jimmy Witherspoon in the course of an interview I taped for my radio show. "Then later I went to California and finished high school there. Then came World War Two and I was in Calcutta, India, in the merchant marine. I sang with Teddy Weatherford's band over there—'Around the Clock,' very risqué tune, Wynonie Harris recorded it, it was all suggestive blues. Before that I was trying to do ballads, which I think most black people were at that time, trying to lose their identity, and I was no exception.

"I don't know why I started singing the blues," Jimmy continued, not a little bit of puzzlement in his voice. "Blues wasn't hardly allowed in my home. My mother was a very religious lady, so was my father. I don't know what made me become a blues singer, I don't until today. I really think had I not been in Calcutta, India, and feeling kind of lonesome from home and just heard Teddy Weatherford, who had been in the East for years and was from Chicago, I think if I hadn't heard him playing Benny Goodman's arrangement of 'Why Don't You Do Right,' I don't think I'd ever have been singing the blues. Seriously.

"The blues is hard to define," Jimmy confessed. "I've often been asked, 'What is the blues?' And I've never been able to do it. A lot of writers, critics have written ridiculous stories. You have to be poor, you have to be black, and you have to be from Arkansas, Tennessee, or Georgia, or Mississippi—which is ridiculous. John Lee Hooker and I are the same age. I'm from Arkansas originally, he's from Mississippi. It's not my fault that he didn't go to school. This doesn't mean that he's a more authentic blues singer. It doesn't mean that several other blues singers—I won't call names—who can't read or write are greater blues singers or inferior or superior blues singers. The blues started with blacks, but it doesn't mean that only blacks can do it. Eric Burdon can sing the blues, Kay Starr can sing the blues, Jack Teagarden, who I knew, could sing the blues, Mose Allison can sing the blues. Everybody has the blues sometime or other. I have the same type limousine the President has—with a chauffeur! But it has nothing to do with me singing the blues!

"Blues is not played on the radio," Jimmy lamented, "and young blacks today are not aware of the blues. I'm ashamed, being a black artist, that not enough young blacks come to see Spoon. It's the media—young blacks only hear rock and roll and top forty. They don't get a chance to hear the blues and they don't go to the library and read up on black culture like they should. I bet you could ask twenty blacks today, 'Who is Jimmy Witherspoon?,' and they can't tell you. They can't tell you that I've played in Carnegie Hall, that I was inducted into the Ebony Hall of Fame, so many things that I've accomplished. There are so many black blues singers and so many young blacks wouldn't know their names. I've seen this happen—young whites in college have turned young blacks on to the blues, and that's beautiful.

"I've been very fortunate to have been around top jazz musicians who responded to my type of singing and who have helped me. Ben Webster and I worked together for three years and I've taken advantage of it, whereas the average blues singer can't work with these men. They don't understand the notes. Bending those notes—you can't learn that at Juilliard. It's something that no music teacher can teach you. The music is just not taught. You show me a jazz musician, if he can't play slow, he's not a jazz musician. Same thing goes for a blues singer. It's an art to be able to sing slow or play slow. A great pianist like Junior Mance, who's the greatest for that, or an organ player by the name of Roy Alexander, who can play so slow a lot of people think, 'Hey, when is he going to get to the end of the chord?' "

So far, we've heard how a great many jazz instrumentalists have learned the craft. It might be instructive now to hear from some jazz singers on their early training.

"I guess I must have been singing ever since I was a kid, singing around the house," Maxine Sullivan recalled in a conversation with me at the Manassas Jazz Festival in Virginia. "I come from a musical family, had a couple of uncles that played instruments, the back-porch type. So I guess it just happened that way. We had a phonograph around the house and I listened to what was popular on records. Course, we're talking a long time ago, you know. I sang 'I'm Forever Blowing Bubbles' in 1919 at the Carnegie Library in Homestead, Pennsylvania, where I'm from. I was eight years old then.

"One of my uncles made him[self] a little crystal set and we used to listen on earphones. And then, of course, radio became a little more available and I used to stay awake all night listening to the remote broadcasts from coast to coast. A lot of the bands came to Pittsburgh and broadcast from the William Penn Hotel every afternoon and I always went to see them—Duke Ellington, McKinney's Cotton Pickers, Claude Hopkins. A lot of those bands are no longer in existence. I listened to Ethel Waters and Mildred Bailey, but I was more or less influenced by the instrumentalists.

"I guess the first jazz I heard was around Homestead. They had a lot of little clubs around there, especially piano players. And, as I said before, my uncle was a musician. He was on the road with a band which Earl Hines was in as far back as 1922, '23, something like that, a band under the direction of a singer by the name of Lois Deppe. My uncle played drums but he didn't stay with the band, he came back to Homestead and he started playing the saxophone. And I used to tag along behind him, try to be the singer with the band he was in then."

Maxine went on to a long and distinguished career that spread over half a century until her death in 1987 and included work with Claude Thornhill's band and the John Kirby sextet in the 1930s and '40s, the World's Greatest Jazz Band in the '70s, and associations with Earl Hines, Bobby Hackett, Bob Wilber, and countless other jazz greats.

"My parents used to play blues records all the time," Ernestine Anderson told me. "John Lee Hooker, Muddy Waters, all the blues greats. In Houston, where I grew up, you turned on the radio and what you got was country and western and gospel. I don't even remember what my first experience with music was. I sort of grew into it. My father sang in a gospel quartet and I used to follow him around, and both my grandparents sang in the Baptist church choir. And they had big bands coming through Houston like Jimmie Lunceford, Billy Eckstine, Erskine Hawkins, and Count Basie."

Ernestine's godmother entered her in a local talent contest when she was twelve years old. "I only knew two songs," she admitted, " 'Sunny Side of the Street' and 'So Long.' The piano player asked me what key did I do these songs in and I just said 'C' for some reason and it was the wrong key. In order to save face I sang around the melody, improvised around the melody, and when I finished one of the musicians told me I was a jazz singer." From that day jazz became her idiom, and the bandleader at her first paying job was trumpeter Russell Jacquet, brother of saxophonist Illinois Jacquet. Ernestine went on to tour with him as well as with Johnny Otis and Lionel Hampton through the 1950s. Ernestine had some lean years, especially during the 1960s, but in the '80s her career began cranking up with overseas gigs, engagements at New York clubs and West Coast venues, and a steady flow of albums.

Chris Connor, who was born in 1927 in Kansas City, had the support of her parents in her musical endeavors but they did not share her interests. "No one in my family was musical except me," Chris insisted, "I was the all-in-one. I used to hibernate and listen to records constantly. During high school I played clarinet for four years. I had first chair and I used to play Artie Shaw's 'Concerto for Clarinet' and all of that. I used to listen to Dorsey, naturally. That was the music of the Swing Era. I heard all the jazz bands going through. There'd be one-nighters at different spots and I'd go out there with my father. And I used to go down to 12th and Troost where the Basie band used to be.

"I'd always wanted to sing. I just set my sights for being a singer because I knew that that's what I wanted to do. It just sort of came naturally. I listened to them all—Billie Holiday, Sarah Vaughan, Peggy Lee, and pop singers like Dinah Shore and Doris Day. There was a college band at the University of Missouri and I used to go over there on weekends and sing with them. They were patterned after the Kenton orchestra and did all Stan's arrangements. That's when I first got my taste of the Kenton band and set my sights on it. I was about seventeen.

"I guess I started singing professionally at nineteen. I came to New York and my first job was with the Claude Thornhill Orchestra. I was with him for about five years and with Herbie Fields for a short time and with Jerry Wald, and then I was with Stan [Kenton] after that for about a year.

"With Thornhill I joined the Snowflakes, the vocal group, and I was singing lead with them. We used to do three or four numbers and then have an intermission and in the second part we'd go on again. You know, sit down, watch the crystal ball go around in the ball room. But with Kenton I was the featured vocalist and I used to have my own twenty-minute stint several times during the night.

"With Claude the whole five years was practically nothing but one-nighters and that really did me in. We'd have two or three hundred miles to make every day on the bus and you'd get to the job, if you were lucky, about six in the evening and you'd have maybe an hour and a half to grab a hamburger at the White Tower and check in a hotel. And I'd have to get my gown ready, put in in the shower [stall] and let it hang out, let all the wrinkles hang out. Then we'd go immediately to the job and we'd be there for two or three hours and then get back on the bus and travel all night, maybe another three hundred miles. You try that for like twelve hours at a time. It's *very* rough. It was a very rough life. It did me in. And I was in my early twenties then. I'll never do it again, I'll tell you that.

"When I finally got with Kenton I couldn't take it much more. I was physically exhausted and I just couldn't take it any more. I just stayed with him nine months. When I wanted to leave the band Stan said, 'Chris, you shouldn't because if you stayed with me another year or two,' I would become more of a national name. He tried to persuade me but that would have meant six years on the road. I said, 'Stan, I just can't take it any more.' He hated to see me go and I hated to go, but I had to. He was just a marvelous man. He encouraged everybody in the band, not just me. He was a great inspiration to work with."

In the 1980s Chris Connor was working mostly as a soloist, but she still sings once in a while with a big band "if the occasion calls for it," as when she was recently on tour in Japan. "I had a sixty-piece orchestra doing a television show. They had about forty strings and six harps. It was unreal, it just about blew my mind to get up there and sing with a symphony orchestra. Usually they have about a fifteen-piece band in some of the nightclubs, and I do my trio things, too."

In the late 1970s Chris Connor came to terms with her dependency on alcohol. She didn't hesitate to talk about it. In fact, she takes considerable pride in having successfully dealt with her alcoholism. "It was a gradual thing, it built up, and I was so sick I didn't know how sick I was. I almost died. It just gradually, progressively, gets worse and you don't know when you cross the invisible line. You can't pinpoint it. I haven't had a drink since and I've never been happier, to tell you the truth. It's a completely different life. In other words, my life is divided into two parts. When I was drinking—it's like that part happened to a completely different person, it really wasn't me. I'm so completely different now. I can remember all the bad things and all the experiences and all that—it's marvelous! Well, if you want to be reborn, I guess that's the word for it."

We earlier alluded to the intensity of jazz activity in Detroit in the 1950s and to some of the artists who came up in that city. The singer Sheila Jordan, a Detroit native born in 1928, recalled for me her early years. Growing up in Pennsylvania and returning to the city of her birth to finish high school, Sheila was exposed to a very different set of cultural experiences, a combination that, in fact, shaped her in a way that is quite unique artistically. It is hard to imagine any other jazz singer bringing off, as Sheila did in a 1962 recording with George Russell, the tour-de-force performance of so unlikely a vehicle for jazz vocal as "You Are My Sunshine." Hear her story and you will understand how she was able to do that.

"I started singing when I was about three years old," Sheila began. "Probably even younger. I *always* sang. I can't remember not singing. We were just singers in my family. My grandmother sang, my grandfather sang, everybody sang. I grew up in Pennsylvania. My grandmother raised me 'til I was of high school age and then I went back to Detroit.

"My musical background was basically . . . country music and the songs of the day, what was on the Hit Parade. The songs of the day were good songs, songs that were made popular by jazz musicians

later. There were these books that gave the lyrics. You heard the songs
on the radio and then you'd go get this book and you'd follow the
lyric. I had a pretty good ear for music so it was pretty easy for me to
learn things fast because if a song was popular, like the rock songs
today, they'd play it over and over and over again. And I went to a
lot of movies and I saw a lot of Fred Astaire movies and a lot of
musicals. The tunes that were in those musicals were the kind of tunes
that are popular in jazz today as a form to start from and then create
from. So, basically, between sort of country and mountain music and
the songs that were popular at that time on the airways is what I got
my materials from, in fact, materials that I even sing today.

"As far as the jazz, I got into jazz when I was about fourteen years
old because I moved back to Detroit, Michigan, to finish my high school
studies with my mother, and that's why I got into jazz. I met Tommy
Flanagan and Barry Harris and Kenny Burrell. We were all high school
kids together, not necessarily the same high school, but we hung out
at the little non-alcoholic clubs and dances where kids got together
and heard the likes of Charlie Parker and Dizzy Gillespie and whoever
the jazz greats were that came to town.

"There wasn't separation of the races," Sheila hastened to explain,
"it just depended on where you lived. You went to the high school in
your area. So let's face it, like the black neighborhood was the black
neighborhood. I mean, there was a lot of racism in Detroit, but it
wasn't segregated in the schools. I started out at Cass Tech High School.
Cass Tech was totally open, as far as the students, I mean, black and
white students went there. I transferred right next door to the High
School of Commerce because I got commercial studies there. Cass Tech
had a good music department and that's where you heard all these
great jazz records on the jukebox downstairs in the hamburger hangout.

"Jazz was flourishing in Detroit when I was in high school, it was
happening. We had a couple of places in Detroit where a lot of the
jazz greats would come and play. You stood outside or went in the
alley, you know, and sometimes they'd work at places where you didn't
have to be twenty-one in order to get in. They were like dances. I got
to hear Bird quite a few times. Wardell Gray was living in Detroit for
a while when I was just out of my teens, old enough to get into bars.
I got to meet Paul Chambers when he was about fifteen years old
because he was at Barry Harris' house one day and Barry said, 'You
gotta hear this kid play!' Our idol at the time was Milt Jackson because
he was with Dizzy Gillespie. Another hero of ours was Hank Jones,
even though he didn't live in Detroit at the time. The Jones brothers
were living in Pontiac and were very prominent after I left. And Betty
Carter was on the scene. She was from Flint and was with Hamp. I
mean, there were so many young people that came up after I moved
to New York. There were a lot of local and resident musicians coming

up. There was a trumpet player called Young Blood Davis. But Tommy Flanagan, Kenny Burrell, and Barris Harris—those were the top people that I remember. It was fantastic!

"My main influence was Charlie Parker. He was the first person that I heard that really moved me to want to hear the music and to want to participate and devote my life to the music. And Bird took me back to all the other great musicians and singers. Through Bird's music I was turned on to Billie Holiday and young Sarah Vaughan and Ella Fitzgerald. Ella Fitzgerald was popular at the time because she'd made that 'How High the Moon' and 'A-Tisket A-Tasket,' you know. And then young Sarah came on the scene and all, and what she was doing was miraculous. But my main influence, as far as singers go, the lady that I *loved*, that I feel most indebted to, is Billie Holiday.

"The thing with me is that when I was growing up there were good songs to sing, there were always good songs to sing, so that helped a lot, and of course, as soon as I got into Charlie Parker I started learning the bebop lines, the new things that Bird was writing, the songs on records that Bird was making, and Dizzy Gillespie and Thelonious Monk and everybody. I mean, these are all the people that I listened to.

"Basically, I listened to instrumentalists more than singers because I wanted to have my own sound. These singers were so great and they were inspiring to me, but I didn't want to try to attempt to imitate them because it was *their* sound. And who could? I mean, really! Who could imitate a Sarah Vaughan or a Billie Holiday? I always had the feeling that to imitate them would just be stealing. I feel you can listen to them and get the feeling of being inspired and still do your own thing.

"But I was such a Charlie Parker freak. Money was very scarce, so when I got money the records I bought were the Charlie Parker 78 records, and if I had a chance I would buy the singers. I had friends I hung out with and we had little jazz groups together and they had a lot of Billie Holiday records. So I got to hear Billie and Sarah through them. But generally speaking, most of my records were of instrumentalists."

We should note here that, except for some piano lessons during high school, Sheila Jordan has had no formal instruction in music. "I've never had any voice training," she pointed out. Yet she finds herself in demand here and abroad as a conductor of vocal performance workshops. She began teaching such a workshop at New York's City College in 1977 and has taught similar sessions in Graz, Austria, in Canada, and at colleges and universities throughout the U.S.

"A lot of times when I go out on tours I do a workshop for a day or for two days. I've been invited to work in Munich this fall and I've been getting invitations from all over the world to come and do a workshop because when I go there and work with the kids it's very successful and they really like it. That's very encouraging. It's amazing and beautiful to see all the young people coming up who are sincerely interested in this music and who, hopefully, will be as dedicated to keeping it alive as myself and many, many other musicians. It's very, very important what's happening in the United States today and in Europe. Now they have jazz courses. When I was growing up there was no such thing as a jazz course. You could not learn the way the kids are able to learn today.

"My course is a performance course. Even though they teach a lot of jazz in the schools, I feel in order to really get to the nitty gritty, you gotta get up there and do it. That's primarily what my course is. I know that it's working 'cause I get the same feeling from teaching as I do when I'm communicating with musicians and creating good music. It's that same kind of a high feeling you get."

Sheila Jordan was maintaining in 1990 a full schedule of performance here and abroad, often appearing in duo with the bassist Harvie Swartz. She has toured internationally with the George Gruntz Concert Jazz Band and has sung in several of Gruntz's jazz operas.

"Like many singers, I was blessed with some very good ears," Carol Sloane told me. "I knew that I could duplicate sounds that I heard." She said that the first musical sounds that she found herself duplicating, recalling church choir practice as a child in Providence, Rhode Island, were sections of Bach and Gounod masses. "Those of us who couldn't read music had to learn them by heart, just as I later learned Jon Hendricks' lyrics to Ellington's 'Cottontail.' "

To this day Carol has not learned to read music, a circumstance that, far from being a hindrance to her, doubtless explains how she can express such extraordinary nuances of emotion with the sort of control and phrasing one customarily hears only from horns.

Growing up in Providence in the late 1940s and '50s, Carol discovered that the fare on FM radio was different from that on AM. "On one wavelength I heard white singers, wonderful white singers, sing great popular music. On FM I heard all these black women who'd had the same kind of training I'd had. They grew up in churches singing a disciplined kind of music. Billie Holiday, Dinah Washington, Ella Fitzgerald, Carmen McRae. I heard Rosemary Clooney sing 'Deep Purple' and then I heard Sarah Vaughan sing it, and I heard two different songs. I decided Sarah's was more interesting to me."

Another singer whose early training included singing in the black church is Andy Bey. "I started playing piano when I was three," Andy informed me. "My mother said I was singing immediately." An early favorite was Louis Jordan's "Caldonia," which, along with other popular numbers of the 1940s, Andy enjoyed singing around the house with his sisters.

Among the experiences that shaped Andy Bey was his exposure, as a twelve-year-old, to the rough-and-tumble Amateur Night at Harlem's Apollo Theatre. The Apollo provided a forum for virtually every notable black singer, dancer, and comedian who came up in the 1930s and '40s. Performance in the Apollo constituted a rite of passage for jazz musicians, and most of the major black big bands, and some of the white bands, appeared there. In the early 1950s, when Andy braved the Apollo's hypercritical audience on Amateur Night and, several years later, when he landed an engagement there, the fare reflected the current popularity of rhythm and blues. Only fourteen years old, Andy worked at the Apollo with the band of r&b pioneer Louis Jordan.

"We would do six or seven shows a day," Andy recalled, but his most vivid memories were of Wednesday's Amateur Night. "If the act was bad, Porto Rico"—stagehand Norman Miller—"would come out and do a funny dance with a funny costume on and, if you were flat, the band would get a signal from him and they'd change the tune and the singer would be confused. He did all kinds of things. He'd make noises, start dancing, make you feel so embarrassed that you'd want to run off. I mean, if you stunk, you knew that you did. By the time he got through with you, you wouldn't want to come back." If this sort of humiliation somehow failed to drive the performer offstage, Andy added, "They'd use a stick to pull the people off."

Notwithstanding the status of New York as the mecca for jazz artists, the idiom does boast players of world class who stay in their home territories and, until the big break comes along, remain virtual unknowns in the international jazz community. Such a story, big break and all, is the career history of pianist-vocalist Ellyn Rucker, who grew up in Des Moines, Iowa, and settled in Denver, Colorado, in the late 1950s upon her marriage at the age of twenty. She was still residing in Denver, where she has been active as a professional jazz musician for three decades, in 1990, but the greater jazz world has come to know of her through several recordings under her own name and through numerous appearances on the European festival circuit. At the time of writing Ellyn was anticipating an imminent New York debut.

"I was part of a musical family," Ellyn began, outlining for me her early years. "My mother played organ in church. She was self-taught, raised on a farm, and they had a lot of musical talent. They were not real well-off and somewhere along the line she learned how to play piano. My grandfather sang in the senior choir and when I came along I was an automatic for the children's choir. I remember loving music as far back as I can remember anything. My family loved good music and it played in our house all the time."

At the age of eight, Ellyn, returning home from a piano recital, informed her mother that she wanted to learn to play the piano and take dancing lessons. "I mean, I was an easy study. Nobody had to drag me there. I *wanted* to study. I *wanted* it." Five years of piano instruction followed, along with the dancing lessons.

It was an older brother, Ross, "who brought jazz into our house. He was maybe thirteen and he had a paper route and he bought himself a string bass and taught himself how to play it. He's the one that started bringing home Charlie Parker records, Jackie McLean, Jackie and Roy Kral, the Kenton things, the Jazz at the Philharmonic series. Ross was the one that I think influenced me the most as far as hearing jazz. Like I said, my mother brought the church faction, that part of the music, into our house and my brother Ross was interested in jazz and I picked up on it very readily. When I heard jazz I thought, 'That's it!' I'm not sure my mother approved at first, but later on she totally realized that this was something bigger than life for us kids. She had to be one of the hippest mothers around.

"And the most marvelous Christmas gift my parents ever came up with was a Zenith floor model console stereo record player, when we all thought all we were getting was some socks and things like that, and they went out and brought this in off the porch. You never saw three more excited kids and that became the focus point of our home life. We were always listening to records and we had kids at the house all the time."

The roster of artists, in addition to those mentioned above, whose 78 rpm records spun on the turntable of that 1940s Zenith included the pianists Art Tatum, Lennie Tristano, and Oscar Peterson. Then there was the thrill of seeing her cultural heroes in person. "All the road acts came through Des Moines on their way to, say, Chicago or Denver, from either of the coasts, and they'd stop and perform at the KRNT Musical Theatre. My parents somehow came up with the money to take us to see stage shows and they also took us to a local ballroom. We'd hear all the dance bands of the era—Wayne King, Guy Lombardo, Tony Pastor—and we heard the Clooney Sisters, Rosemary and Betty, before they were anybody, and we heard Stan Kenton and Woody Herman's Third Herd. I saw the Duke Ellington and Count Basie bands on two or three different occasions, the Oscar Peterson Trio with Herb Ellis, Gene Ammons, Teddi King, George Shearing, and Jazz at the Philharmonic with Lester Young, Illinois Jacquet, and Flip Phillips. I'll never forget seeing Sarah Vaughan. I was probably sixteen or seventeen and she just blew me away with her presence, her persona, and her wonderful singing."

Another singer-pianist, Chicago native Judy Roberts, enjoyed even stronger encouragement from family. "My father, whose name is Bob Loewy, is a terrific guitar player and singer," Judy explained. "He grew up in the Big Band Era, did arranging for Fletcher Henderson, played with Muggsy Spanier, Israel Crosby, and those kind of people, and he knows a million tunes and has a great ear. So I was born into a complete musical settting and in lieu of material things, which we definitely didn't have, we just sat around and played piano and guitar all day as recreation.

"By the time I was eight years old I must have known a couple thousand songs by osmosis and that is why, to this day, I don't hardly read at all. For a game my father would say, 'Okay, let's see if you can figure out the chords to 'Tenderly,' and I would figure them out by ear. I remember pictures of me sitting on the piano with my feet on the keys, like about eight months old or something, and my dad playing. In the middle of a gig, if I get hung up on a tune, I call my parents and one of them can sing the entire song to me. My mother can sing the bridge or any part of any song. She just has a retentive memory.

"In high school I was already a weirdo," Judy recalled, "the class beatnik. I was listening to André Previn and everybody else was listening to the pop garbage of the day. So I was an outcast, a poor, Jewish outcast. I was dating a jazz bass player and he was working at a coffeehouse in the suburbs. One night I went out there and the piano player didn't show up, so guess who sat in. The owner said, 'Hey, this is great! A girl piano player!' I was hired, catapulted into a coffeehouse job at age fifteen. People say, 'Weren't you discriminated against as a woman?' 'No, just the opposite,' I answer. 'The gig was dependent on pleasing this owner who decided that a girl piano player was a commerical commodity.' "

Judy's debut as a singer came about in an equally inadvertent manner. She was working a 9 p.m. to 5 a.m. job in a sleazy Chicago bar in the 1960s. "This guy came up to me and said, 'Do you sing "Fascination"?' I said, 'Gee, I'm sorry, Sir, I don't sing.' The bartender, who was a friend of this guy, took this gun out and slapped it down on the bar and said, 'Listen! Sing "Fascination"!' So of course I sang it. Later I said to myself, 'God damn, if I can sing "Fascination," I can sing all these songs I've been hearing Jackie and Roy do and Nancy Wilson do and Annie Ross do.' And the next night I was a singer."

Perhaps the most unique, even innovative, jazz singer to emerge in the 1980s, a cappella performer Bobby McFerrin is truly in a class by himself. Lyrics, scat, vocalese, simulated horn lines, string effects, and body percussion combine for a one-man show unlike any that has ever graced a bandstand. For some selections, for example "Take the 'A' Train," he coaches an audience in riff patterns and then sings an improvised solo over their big-band ensemble sound.

New York-born Bobby McFerrin was raised to the household sounds of Bach, Verdi, Mozart, and other European classics; his parents were opera singers. It wasn't until 1970, when he was twenty, that he was turned on to jazz.

"I met this girl in college who really liked jazz," Bobby recalled for me. "We went on a date to see Miles Davis. Keith Jarrett was on piano, Airto Moreira was on percussion, Jack DeJohnette on drums, Gary Bartz, saxophone, Michael Henderson on bass. You can imagine the force and the effect that it had on me. It was incredible. I think right from that moment my life was different somehow. It turned me around and I really became serious about finding out the relationship I was going to have with jazz."

Bobby McFerrin had studied piano at the Juilliard School and at Sacramento State College, and in 1977 he was pianist for the University of Utah Dance Department. "I was looking for something else to do," Bobby explained, "because as much as I really enjoyed accompanying dance classes, I was looking for something else to do, something that was a little bit more satisfying, really. I was in the midst of sort of self-introspection. The decision to sing came up and it was very strong leading me and I followed it obediently.

"After I started singing I was going through different kinds of images of what kind of singer I thought I was or would like to be and this recurring imaginative sort of picture kept coming up. I would see myself on stage by myself. I couldn't really understand what that was but I was thinking I had always been attracted to solo artists and one-man shows ever since I was a kid and I really dug them. I remember when I was twelve years old and I first heard the word a cappella, I, thought it was such a beautiful word, you know, so all those things, I think, culminated in this idea of solo singing.

"First it was the idea of what I wanted to do and then I had to devise a technique that would make it work, and that's what took a long time, a lot of trial and error, and teaching myself how to use space effectively and how to breathe rhythmically and how to sing intervalically, all those things combined. Staying in tune and implying harmony with melody had to be worked out some way. The only way I could really detect that it was coming across would be when people would come up to me after a performance—and when I first started doing this I would do only one or two solo things in an evening—and they would say, 'I could hear everything, I didn't really have the sense that anything was missing because I felt the rhythm and I could hear the harmony and I could distinguish the melody as you were singing.' That kept me going, that was enough incentive, because I knew I was on the right track when I got that kind of feedback.

"I picture concert halls like a blank canvas and in a lot of ways I think of myself as a painter who works with sound colors," Bobby said, making use of a very effective metaphor. "It's very challenging, stimulating, and quite exciting to be up there by yourself. You're totally self-sufficient, self-reliant, and you just have to keep pulling things out of the hat, you know, digging deep down inside you for ideas. The hardest part, though, when you're improvising and you're trying to *find* an idea, is simply waiting for something else to happen without trying to affect it too much, going to those places that you know work. The ulitmate experience is waiting for something to happen and it happens on its own, or you hear something being affected by the rhythm or whatever you're doing and you begin to hear something else. It's waiting for that moment that is the hardest part for myself and, I'm sure, for the listeners a lot of times because they're just sitting and waiting and wondering what I'm going to do. So there's usually quite a bit of tension in concert, not a negative tension, but the creative kind, that stimulating, kind of on-the-edge feeling."

In addition to the help his parents gave him in "finding the right voice," Bobby cites Keith Jarrett as a model for solo performance. "He helped fan the flames of the solo idea. Charles Ives was a great influence, too, because he fueled the idea that a song has a life all its own, it has a certain life, whether it was written down or not, and he helped me to just sort of go with my imagination, give my imagination full range, because it could do anything it wanted to do."

In the course of talking with many jazz vocalists over the years I found myself curious as to what role the singer played in the song. Here are some of the responses I got.

"I'm a conduit, I interpret," Rosemary Clooney told me. "Most of all I look for a word that means something to me, appeals to me, reminds me of something, a memory or some kind of feeling, as opposed to perhaps the way I started, when I could sing something a little nonsensical like 'Come On-A-My House.' " The reference was to the song that gave Rosemary her first big success and made her a household name throughout the land within several months of its release in 1951.

"Singing is acting, really," Shirley Horn explained. "I try to get inside, try to step inside a song. If the lyric is heavy, I try to close the curtain, try to close the audience out, because it's a little distracting. I don't like to work in a place where there are lights flickering all over because if I see people picking their noses, it disturbs me. Yeah, I try to step inside, away from the madding crowd."

Shirley is one of a handful of jazz singers who accompany themselves. In truth, she is every bit as fine a pianist as any number of jazz pianists out there who never sing a word. "I can play what I want for me," she observed. "I don't have to suffer not knowing what a pianist is going to play for me. I just suffer with myself," she added, laughing, and then confided, "You're supposed to have the rags-to-riches story, right? I didn't have that story. I didn't have to pound the pavement, it sort of came easy to me. I had good people in my corner. I mean, I didn't have to perform if I didn't want to." Shirley paused, gathering the thought together much as she might fill in with a few bars from the keyboard before going on to the next verse of a song. "But I *have* to perform. There's something that drives me. If I go too long without performing, I get crazy."

Barbara Lea, one of the ablest interpreters of the work of George Gershwin, Jerome Kern, Hoagy Carmichael, and others who have contributed to the body of classic American song, has also worked in the theater. She talked about the two experiences. "I was so ill at ease when I was performing"—as a singer, that is—"in public that I started studying acting to overcome that stage fright and I fell in love with the theater. I found that I didn't get nearly as scared acting because everything was given to you. First of all, the whole script was given to you, the part was given to you, the costumes, the direction, and the other people to relate with. And everything got rehearsed and it was all set up. I'm not doing anything as myself, I'm doing everything as that other person. No matter how close to me that other person may be, it's not like me making a personal statement, it's that other person making a personal statement and I'm putting my emotions and my perceptions into it. It's like loaning myself," she chuckled, "to another person.

"What I try to do when I'm singing is put myself completely into the song. I try to deliver the lyric, deliver the melody, deliver the song. Instead of trying to make myself look good, I try to make the song look good. When I hear singers that make me think about *them* and forget about the *song*, I get real mad at them. And it doesn't mean you have to be a slave to what's written, but just don't get up there and show me how cute you are. It seems to me that their intention is to get you to admire them. That's something I learned from acting. You don't get up there and play sad or play happy. You get up there and try to take the action that the character is doing. You don't get up there and try to cry. You get up there and try *not* to cry, which is what the character is doing. So I don't get up there and try to show you how much I'm waiting for the man I love. I get up there and try to *wait* for the man I love, and just let you in on it.

"I try to give the kind of performance where there will be a work of art there that is the meeting ground for me and the audience, like a painting. The painting is outside the artist."

Melissa Berman, a young singer in Washington, D.C., confirmed that Barbara Lea's approach appeals to other performers. "My approach to my music is as a musician and as an actor," Melissa said. "When I work on a song I'm working on it as a theatrical piece as well, as if it were a monologue or a role in a show. I approach my roles in the theater as if they were songs sometimes, too. There are certain rhythms that characters have, their own musical pattern. In some songs I may bleed a little," Melissa confessed, "but I haven't done that for theatrics, I've done it because that's where the emotion has led me, where the song has led me. Acting is really *re*-acting. Bertolt Brecht worked with his actors to push the pain down. I use that a lot when I'm singing. You don't want to shoot your wad in one song."

Little Jimmy Scott, who began singing in church choirs in Cleveland, Ohio, as a youngster in the 1930s, joined the Lionel Hampton band in 1949 at the age of twenty, and went on to work with Charlie Parker and many others, will conclude this compilation of singers on their approach to the interpretation of song. "It comes from self-expression, really," Jimmy explained. "We were taught to sing with sincerity even when we were youngsters, so you developed from that natural thing to expressing the style or story that was involved in the song. I say that it should have meaning to the singer for the singer to be able to express it to the public."

Of all the many sources of song that jazz singers draw from, the so-called American song book constitutes a core of the repertoire of many vocalists. Already in this chapter we have heard several singers allude to this circumstance. Here are some more precise assessments of the value of that body of song created by the likes of Irving Berlin, Jerome Kern, the Gershwins, Harold Arlen, Johnny Mercer, and others.

"I love it, I love the music," was Washington, D.C.–based vocalist Deater O'Neill's enthusiastic clarification of why she sings the old songs. "That's the bottom line. American popular music is a priceless and timeless legacy, and I'm proud to help assure that this national treasure is widely shared."

"You can't get much better than that," insisted singer Carol Fredette "and we're trying to maintain it and sustain it so that it will endure. Because if those of us who are around don't keep it going, it's going to become extinct. It's like you've got to believe in it with your whole heart and soul. There are people out there who feel as we do for the kind of music that the world wrote when there were Kerns around and Gershwins and Cole Porters. When I go into clubs, the younger people *love* it and they really want to hear it."

"I love doing new things," Carmen McRae said when I steered our interview toward her choice of new materials for her ever-expanding repertoire. "Not that I don't love Gershwin and Cole Porter. I love all their tunes and it's the *basis* of my career, what I've done most of my life, but it is nice to have some new music around. Naturally, *most* of the contemporary tunes written, I can't do or they don't suit me or I don't like them. The ones that I do find interesting is because, first of all, I do love lyrics above everything. Sometimes the new songs are not as good lyrically as they are melodically. There's ones that I hear that I can't understand the lyrics because they don't come over too clearly on the records and, if I like the melody and I hear a couple of words I like, I'll go and seek out the music and find out if the whole thing is worthwhile.

"Now when I do go about finding one of them, I like to find something that *I* personally like and personally think that I can do, do it well enough to entertain the people who have been used to listening to me do nothing but Cole Porter and George Gershwin. I find that they like it.

"We have some great contemporary writers and they do have some great lyrics, some of them, and those are the ones that I choose. Other than that, my pianist will come to me and say, 'Hey there's a new song and I think you'd like it.' Oh, and my very best friend has two daughters who are *hip* to the contemporary tunes and they know what I do and they always say, 'Oh, Aunt Carmen can do this tune,' and that's how I get a lot of material also."

Then there are the singers who write some of their own songs. I put the question to two of the very best: How do their creations see the light of day?

"I have to be pressured into beginning to write it, number one," Dave Frishberg confessed. "And number two, I think of a title. That's what comes first, and then I usually work from the title and let the song happen that way. I'll *make* it happen, to justify that title. The last couple of days I've been stumped because I can't really get a title. I know exactly the tone of the song I have to write—it's for a show I'm writing—and I know exactly what it's supposed to be about and who the guy is that's singing it, but I can't begin writing it because I don't have the title. As soon as I get the title it'll be more like a puzzle I have to complete. As soon as I get that title, that will trigger a rhythmic impulse and usually the outlines of the melody are in my head as I'm writing the lyrics. I don't write the lyrics first and then set it to music. I usually have some kind of a musical idea that I'm humming along as I'm writing those lyrics."

The reader might enjoy imagining the process which Dave describes, applying it to several song titles that this brilliant composer-lyricist has started with. There are, for example, "Blizzard of Lies," a catalogue of the mendacities we routinely commit ("The check is in the mail," "I'll always love you, Dear," etc.) "Van Lingle Mungo," a roster of Dave's baseball heroes, and "Z's," an homage to shuteye.

"Strangely enough, it comes to me complete," Jon Hendricks said, explaining how he so artfully tailors lyrics to the musical fabric of Count Basie, Charlie Parker, Duke Ellington, and others. Jon, who first came to prominence in 1958 as a member of one of the most popular jazz vocal groups of all times, Lambert, Hendricks & Ross, has been dubbed the Poet Laureate of Jazz. "I just have some strange kind of gift. It just comes into my mind all complete. I just can't get the words down fast enough sometimes. It's kind of weird."

The jazz singer's art is not easily defined. Over the years I have asked not a few jazz vocalists to, if not define the idiom in so many words, at least provide me with a description of what they see themselves as doing. The two very best such accounts were given me by artists whom some consider to have been the premier jazz singers performing in the 1980s, Betty Carter and Mark Murphy. It is not surprising that the two went about answering my question in different ways, and both digressed, consequently widening and enlivening the discussion.

"Oddly enough," Mark began, "this television producer was showing me around Brno, Czechoslovakia, and he really put his finger on it. He said, 'You know, you sing not the melody but the harmony.' I really am a harmonic singer because when I scat I'm very sensitive to the piano player I'm working with. And I have to tell you, because of what I call 'harmonic trauma,' most piano players close my ears. But a few of them really open them up, when we like the same chords, and then I can scat for hours.

"The reason I like Jack Kerouac is because I love his words. I like the flow of words and the way words sound. So it's got to be a very astute combination and maybe that's why my favorite composer might be Cole Porter with Harold Arlen and Johnny Mercer as close runners-up. With Cole Porter the harmonies are there and the fantastic word thing is there, like the lyrics of 'In the Still of the Night.'

"See, I learned so much about vocal arranging from being a Peggy Lee freak for so many years. I learned all the possibilities you can do with a trio and ways of keeping an audience's attention by dropping instruments away, suddenly being with just one instrument, suddenly just with drums, using all the aspects of the drums, using percussive elements out of the piano."

Mark offered an illustration of the foregoing, a recorded collaboration with alto saxophonist Richie Cole on the Oscar Hammerstein/Jerome Kern classic "All the Things You Are." "There's a short vocalese intro and then we state the song—I think you've got to do that first—and then Richie and I do an improvisation, and then someone else plays and everything stops and I go back to the verse and do that strictly a cappella and out of tempo and then bring in a completely different tempo. I think I gave everything I could think of to that particular song, which is probably one of the best songs for harmonic possibilities. I don't know how they do it nowadays, do you, when they've got two changes for a whole tune, or three? How"—his voice was skepticism personified—"do they keep thinking of ideas? Tell me.

"My mother and father met in the Methodist choir. My grandmother was playing the organ and when she retired her daughter-in-law took over and she's still playing it. My father was one of those renaissance men—his profession was lawyer—and both he and my mother had glorious voices, just under the level of, say, opera voices. Sunday afternoon it was listen to the opera and afterwards doing all the choir work. It was the best musical training in the world. And then my uncle, the youngest member of my mother's family, placed in front of me one day a record of Art Tatum's 'Humoresque,' which if you remember he played sort of semi-classical first"—Mark hummed it in a stately manner—"and then he swings it in the middle. It absolutely fascinated me and I've been hooked ever since. That was when I was about fourteen."

Mark asserted that "right from the beginning" he was listening to "the great ballads and the great jump tunes," and from discussions with fellow singers he said that he had concluded that this was a common experience of vocalists of his generation. Mark was born in 1932 in Syracuse, New York. "Our ears opened up in the right way. Probably this would happen with writers and painters, too. You know, you're born and then you're influenced and then early, late, or middle you start developing, and there is some kind of strange parallel of taste so that you all start selecting the Duke Ellington ballads, the best of the screen-writing songs, like 'Prelude to a Kiss' and 'Star Eyes,' all the great songs. I suppose my *very* first influence was Nat Cole and then June Christy and then Ella Fitzgerald. So I learned to sing in a very direct on-the-beat swing style. Later on I went through a heavy voice period of memorizing Sarah Vaughan and Billy Eckstine and then I didn't listen to anyone but Peggy Lee for a while.

"Then the bop era came in and I can remember standing in a music store downtown and these bebop single records would come in and the guys would be in there listening to George Shearing wiz through a thousand notes." As for big bands, it was "mainly Kenton and Woody Herman." A little later, "that *Birth of the Cool* album must have made quite an impression on me because out of that came 'Boplicity' on the Kerouac album and this lyric I wrote for 'Godchild.'

"I probably will never get to know what it feels like to sell a million records, but people do buy my records and keep them for *years*. I've become part of their lives and they're constantly championing me to their friends and it slowly spreads that way. I must do something when I record that becomes very personal. I don't know what it is, it's some sort of communication. You know, you send out your own vibrations and people who like mine pick them up. So this thing travels by word of mouth and by good people like yourself who write for jazz.

"I can be in an airplane with Gladys Knight and the Pips. Gladys is in first class and I'm back in second with the Pips. And each Pip makes more money and is better known than I am. But I'm on the plane, I'm working."

Nor is Betty Carter all that well known apart from the international jazz community, although in 1988 her Thanksgiving appearance on *The Cosby Show* in the role of Amanda Woods, a jazz singer, certainly lent her, albeit ephemerally, some wider fame. I began the interview by asking her to account, if she could, for the individuality, some would say idiosyncrasy, of her singing style.

"Well, because I'm a different person, I think that's the real reason for it. I don't talk like anyone else, I have a different pitch, too. When I say, 'Hello,' it's different from anybody else's, so I think that has a lot to do with it. That's the reason why you get different sounds, 'cause everybody *is* an individual.

"When I came up in the business years ago, in the late '40s and early '50s, it was like that's the only way I could survive. I couldn't get anywhere sounding like Sarah Vaughan or Ella Fitzgerald. I *had* to be myself, that was the code in those days, to be yourself, to *be* an individual, in the '50s. It was easier to be an individual then than it is now.

"My influence was really the music itself. The new music that came up was very important to me, bebop, that whole world was new to me and I loved it. What Sarah Vaughan and Ella Fitzgerald did was open up avenues, they opened up avenues for singers to travel in, if they cared to. In other words, you could do a little more with a melody than years before when you had to sing the melody straight. Ella and Sarah introduced the melodies in different forms. With them opening the doors and the new music the musicians were playing, if you had any sort of creative ability, you could develop something on your own, if you *listened* to the music.

"I was listening to anything that was on the radio, you know, at that time. I didn't have a record player. So we heard everything—the Hit Parade, Paul Whiteman, we were exposed to all kinds of music. But my brother-in-law brought some records of Duke Ellington to Detroit when I was about eleven or twelve, which started a whole new scope for me. And then, around sixteen, seventeen, bebop came to Detroit."

I couldn't resist asking her, how did bebop hit her?

"Strong! Ha, ha!"

We moved on to a discussion of the jazz scene at the time of the interview (1977).

"The jazz scene today is not what people think it is, because we're putting the jazz tag on a lot of different kinds of music. We're throwing the word around. In other words, we're saying to the audience, certain people in the business are jazz people, yet jazz is not supposed to be the kind of music you're supposed to get out there and work for to make money or to get a record date. You can't just go into the studio and do jazz, you gotta do jazz-rock, they say, or jazz-soul or jazz-whatever. So the young kids today don't have a real distinctive picture of what jazz really is, 'cause you can't just put a jazz label on somebody and say they're jazz.

"It's a long, hard struggle to become a jazz artist, it's not an overnight thing that can happen like a lot of hit records happen today. You have to pay dues, you have to spend some time learning the music. It's a learning craft.

"During the jazz scene, the real jazz scene, there were more individuals that came up through that era. We could pick 'em out by the touches on the piano, we knew who was playing the piano by their own individual style. This doesn't happen today. You have to sound like Elton John, you have to sound like Aretha Franklin, or somebody else to make hit records. The individual isn't there.

"It's a cover-up, it's like, 'I'm doing this because I really want to make the money,' which is the main thing, and, 'I want to reach a wider audience,' which is another thing that they use as excuses for watering down the music.

"There's so much money to be made by the manufacturers of these new electronic instruments. Everytime you go to a music store there's always a *new instrument* coming up, somebody has invented something *brand new* and it's on the market, it gets on the market so fast. I don't see any let-up on the artist being exposed to electronic instruments. And the human element is gone the very moment that you put your hands on the electronic instrument because you can't get your *own* softness, it's a *dial* softness, you can't create your *own* emphatic dominance, it's a *dial* thing, it's tricks and things like that.

"Jazz is spontaneous, it's on the job that it happens. I mean, it's not something you set up and say, 'Okay, here we go,' and do the same thing each show, you know, every riff, everything is *set,* it's potted. That isn't what jazz is all about. You have your theme, and after your theme, in every chorus it's different. I mean, you *try* to make it different, because this makes you learn.

"But if you're set in one way of doing that music every set, and it's the same way, and you yourself as an individual can't change that, then I don't think you're qualified to become a jazz singer. You're not qualified to *be* a jazz singer unless you can improvise *on* the spot, *at* the moment, under *all* kinds of circumstances. Maybe the piano player goofs or the changes are wrong or you start off in the wrong key or something happens—you're supposed to be able to handle that without falling apart. Your musicianship is supposed to help you get over that little tragedy that just happened a few minutes ago.

"But today the skills are not really equipped to handle that, because of the times. And it's not their fault, it's just the way things are today. You're produced, they're telling you what to do, how to do it, how to sing. It's all about money."

9

Jazz
Around
the World

In Europe I'm considered an artist and here I'm just another drunken musician, more or less.

Wild Bill Davison

In Japan, more than anywhere else—it's incredible—they know in what years, in what months, you did such and such a thing.

John Coates, Jr.

Jazz is the only original American art form and it's so strange that everybody in the world knows that except the citizens of the United States of America. Isn't that incongruous? I mean, it's unbelievable!

Dexter Gordon

Of all the developments that have directly affected the ability of jazz artists to earn a living—to survive as active professionals in many cases—none looms as large as the veritable explosion worldwide of the interest, even obsession, with which the idiom is greeted in Europe, Japan, and other parts of the globe.

I spent several hours with Dexter Gordon the day following his quartet's opening night at the Showboat in Silver Spring, Maryland.

He had recently returned to this country from a fifteen-year residence in Europe.

"I really didn't have the intention of staying over there," Dexter assured me, recalling the circumstances of his very first visit to Europe in 1962. "I figured about three months and seeing what I could see. In fact, I still had my apartment in New York. I had sublet it to a friend. So pretty soon I started getting letters, 'Hey, what're you going to do, when you coming back?' I said, 'Well, pretty soon.' You know, vague answers."

I asked Dexter to try to recall the factors that had conspired to make an expatriate of him for a decade and a half. After all, I pointed out, he had not returned to his homeland, except briefly for gigs and record dates on a few occasions, until earlier that year.

"They have a deep appreciation for artists over there," Dexter pointed out. "They give you a lot of respect, they don't base their opinions and so forth on material things. If you're an artist, they accept that fact and they show a lot of respect for that. It's a nice feeling, you know, being an American, where here that's not understood or heard of."

Dexter also spoke of the musicians he worked with during his residence in Copenhagen and on frequent tours of the continent. "When I first went over there it was very difficult to find competent professional musicians because all of the people over there that were playing jazz had other means of livelihood. They were doctors, lawyers, engineers, salesman, et cetera. Through the years, with so many American musicians being in residence and touring through Europe, the quality of the musicianship has improved one hundred percent all over Europe. There is a class of regular professional jazz musicians in Europe today, which I'm very happy about because I work with them often. It takes a lot of weight off me because I work with different groups throughout the different countries and so it helped a lot."

I got together with Dexter again upon the occasion of one of his week-long engagements at Washington, D.C.'s Blues Alley. We pursued some of the same themes. Dexter had done quite a bit of traveling in the several years since we had last talked.

"Since I spoke to you last we've played in Mexico: in Mexico City, in Chihuahua city, and in Puebla city. In Chihuahua city a young man—he might have been the only one in the city that knew anything about jazz—he said to me, 'Oh, Seen-yor Gor-r-don-n, you know, thees ees the fur-r-rst jazz band to ever-r-r come to Chihuahua city!' That accent and everything made it much more interesting, huh? But what I'm saying is, I'm sure it was true, you know, Chihuahua city in the state of Chihuahua next to Texas. It was from there that Pancho Villa would make his raids into the border towns in Texas. Very historic, right?" Dexter chuckled over this final observation.

"It was the same circumstance, I think, when we played in Athens.

That was the first time I'd ever been in Athens, period, and it was a lovely experience because the people were very responsive. It was a nice theater, standing room only, a line outside. I mean, it was packed. They really came out.

"But you see, what's so interesting about something like this is that these people were not necessarily jazz fans. Many of them, of course, had some knowledge of jazz as to different styles and different musicians and so forth, but the majority of them you would never say they were jazz fans. They came to hear the concert because they love music. I think that's the first time they'd ever had a bebop concert in Athens. I think they'd heard of maybe Basie, Ella Fitz, maybe Sarah, somebody big, maybe Dizzy, but that's a little different flavor than what I do. I mean, they have a deep, rich musical culture, but so far as jazz was, they didn't know if we were playing bebop or Dixieland or swing. I didn't really know what to expect.

"When you play in a group and somebody takes a solo, you know, there's some applause, whether light or loud or long or whatever, but these audiences, they don't do that, because it's only in jazz that people do that. So when we played and after the solos, just nothing. I said, 'Wow, we're really bombing!' But after we finished the tune— 'Bravo, bravo!' They loved it, they loved the music!

"And that's so interesting, because they weren't a jazz audience, they have no jazz heritage. You know, jazz is the only original American art form and it's so strange that everybody in the world knows that except the citizens of the United States of America. Isn't that incongruous? I mean, it's unbelievable! When you go out of the country, that's the thing they dig about America, is jazz. They're not too concerned about a Ford or a Chrysler or something. I mean, this is what America is known for all over the world."

The trumpeter Benny Bailey told me between sets at Washington, D.C.'s One Step Down of some of his European experiences. "I was over there with Dizzy Gillespie's orchestra in 1948 and I liked it, I liked the atmosphere, I liked the reception we got over there. It was really fantastic, everybody loved us because the music was new, you know, bebop music. It seemed like everything was timed perfectly because after the war here comes this fresh music and probably any other time they might not have accepted a new music that readily. But I think at that time they would have accepted anything from America that was new because the whole trend was on a new beginning.

"They just really, oh, welcomed it with open arms. In Sweden the musicians even began playing it. We got crowds, which was something because they didn't know anything about the music at all. At that time guys like Coleman Hawkins and Roy Eldridge was known over there, more swing type of music, they were pretty well known, and Dixieland, they played Dixieland.

"Ever since I came back with that band," Benny said of his return to the U.S. from the six weeks or so tour of Sweden, Belgium, and France, "I had already made up my mind that I wanted to go back the first chance I got. So I got a chance finally in 1953 with a sort of revue, a pop band actually, a show called 'Harlem Melody' with a couple of dancers, things like that."

Benny settled in Stockholm after the tour ended and remained there for most of the remainder of the decade, after which he relocated in Germany. "Sweden was very good for jazz. There was three or four excellent musicians there and excellent arrangers that had been listening to a lot of guys here. Mostly they modeled themselves on the musicians over here more or less, but they had knowledge to go with it. We had a group of about seven pieces, a very good group, all Swedish.

"In Sweden at the time everybody spoke English and they tried to live like Americans. They listened to American music, watched American movies, and everything. We traveled on weekends, as far as six and seven hundred miles in a car, and then come back to Stockholm. That was pretty rough in wintertime over icy roads. But I got a chance to see a lot of the country, all the small towns, the North Pole, and I learned the language. Every summer we played a tour that went out to the open parks all through Sweden and this was funded by the government. We played dance music which was jazz, and at that time people would dance to jazz music, to bebop.

"It was a very, very good time, a very welcome atmosphere. There were reviews in all the newspapers. I could play the kind of music that I wanted to play. They were really on top of it."

The flugelhorn and, formerly, trumpet player Art Farmer settled in Vienna, Austria, in 1968. During a visit to this country he came to Charlie's Georgetown in Washington, D.C., and we talked about, among other things, his reasons for moving to Europe.

"I got a job offer with Austrian radio broadcasting where I would be under contract for nine months out of the year and I took it because I only needed to be there ten days out of the month. I have a quintet that's based in Vienna composed of mostly Viennese musicians, although there's an [American] bass player named Reggie Johnson who lives in Switzerland who comes and plays with me. We make a short tour two or three times a year. Other than that my work consists of general freelance from place to place—festivals, clubs, concerts, TV engagements, usually with local musicians. Many of the concerts are filmed, many, many. In fact, if it wasn't for the radio and TV participation, there wouldn't be as many concerts. This year there's supposed to be a movie being made of the life of Lester Young and Bud Powell and I'm supposed to be in that. I'll be playing the part of a musician of those times.

"It's kind of hard to explain," Art confessed, responding to my query,

How are jazz musicians treated in Europe? "I usually say there's more exposure on radio and TV. Generally speaking, there are only about two or three stations in each city. Like here you might have twenty or thirty. These things are controlled by the government and people pay taxes for radio and TV. So if there's a group of the population, say if there's a small group, maybe three percent of the population wants to hear a certain kind of music, well, they have a right to hear it and their wishes have to be dealt with. Here they can be completely ignored, the people here who want to hear jazz are completely ignored when you get to NBC and CBS, et cetera. Over there they have to be considered.

"The funny thing, the most ironic thing about being over there is that I play concerts in small towns where the population might be ten thousand and I can go into a place like that and fill up a concert hall of maybe five or six hundred seats. The thing that amazes me is that the people come there and they know me. But if I go to some small place here, where the population is ten thousand—it could even be much more than that—it would be hard to find someone in the *whole town* who knew anything about me."

Cornetist Wild Bill Davison, whom we heard from in the chapter on Chicago, moved to Copenhagen in the mid-1970s and continued to use that as a home base into the early '80s. We had several substantial interviews during this period, getting together when he came to Washington for gigs.

"In Europe I'm considered an artist and here I'm just another drunken musician, I think, more or less," Bill confided to me after several years' residence aboard. "And the fact that I have made so many records and so many musicians have tried to copy the styles that we set here, I became sort of a little tin god to some of those people. It's really embarrassing sometimes the way I am treated. Musicians are taking my coat off and getting a chair ready and will I have something to drink and is there anything we can do and all this. And of course I was not used to that here.

"When I first went over I had a thirty-five-day tour, which went into three years, and I didn't want to leave and I don't want to leave now. I don't want to come back here to live and work as a musician. I like to come back and do a tour like I did this time. It's just unbelievable!

"Of course, when I first started on this tour of Denmark I couldn't *believe* what I was seeing because the music I play and have played all my life is the popular music there and so at my age—I was sixty-nine—I've got a brand new ballgame going. I can do what I like, I play like I like. They dance to my kind of music, every *kid* in Denmark knows me. I always liked that one expression, 'It's nice to be a big frog in a little pond,' you know that thing. And it's nice to be respected and know that you're really wanted as a musician. I'm not fighting for an existence, it's there for me. It's just like when I started in the '20s, it's just like starting a whole new life again. I don't know how long I'll last at this but I intend to try to last a few years." Bill was still touring Europe, Japan, and Australia in 1989 at age eighty-three, leaving this world in November of that year.

In the first two years of Bill's residence abroad he recorded seventeen albums, including one each in Poland and Czechoslovakia. He expressed high regard for the musicianship of European traditional-style jazz players. A life-story film featuring him had recently aired on Danish television. It was in color, from baby pictures to the present, and included scenes of him performing with different bands, talking about his career, and "a bit of acting, which of course I'd never done, walking down the street and going into different clubs.

"I'm traveling almost all the time. There's no such thing as a steady job. You're constantly on the move. I told them to get me a night off occasionally because it's really rough on me. I'm going on trains and planes and hydrofoils and submarines and everything. When I tour Denmark there's about twenty-five or thirty clubs that I do at one time, every night a different club. But it's a very small country so the jumps aren't very large, mostly by bus, and you can go on the train any place there. I just did forty-four one-nighters with the Dutch Swing College Band in Germany. There are some *excellent* jazz bands in Italy, most of them consisting of doctors and lawyers and such. They're not professional, they play for their own amazement. A lot of American jazz musicians have come to Denmark to live like I have. I played a radio concert with Dexter Gordon not too long ago. I'm working constantly there, so much that I sometimes turn work down. There's always something to do there. All the guys have sort of taken me in and they say that I no longer can leave Denmark, I must stay there because now I belong to them and all this kind of stuff," Bill laughed, "which is very flattering and nice.

"In 1957 Humphrey Lyttelton's band went with the Eddie Condon band and we toured all of Europe and Scotland. It was a pretty hectic tour, a pretty wild scene all the way through. But that's how I got going to Europe on my own and I've been going from that time to now almost every year. It's funny that traditional jazz in Britain, from what I can gather from the guys I saw at the festival this year at Nice, has cooled off considerably and there isn't as many jobs or jazz clubs for it as there used to be.

"But now Denmark and Sweden and Holland are just unbelievable. It seems like the whole thing has moved in that direction and all the English bands play that territory, too. I'll just give you an idea of it. Denmark is probably the size of the state of New Jersey and they have ninety jazz clubs and about eighty-five percent are actually financed by the government. They might take an old warehouse and rebuild it and it might have three floors. On the first floor they'll have traditional jazz, on the second they might have folk music, and on the top floor rock. And they have a restaurant and all the kids can go there and they can drink cheap and eat cheap and it's really very reasonable to go and spend a night and hear all these bands if they want to. And touring bands come through, and guys like me, for instance, being featured with the local bands. Actually, there's more work than bands over there can really take care of, that's how much work there is, just unbelievable.

"You see *no* older people or middle-aged people out on the dance floor and of course the guys in the bands all dress like the rock musicians here. The first night that I went to a place to play I looked at the band and I thought, 'My god, I hope they haven't got me mixed up with a rock guy,' figuring that these guys have gotta be a rock band. It wasn't at all, it was a *very* fine band.

"In Holland and places that you'd least expect it, say, Norway, you'd see this town and you'd wonder where the people would come from. In Norway I played in one place surrounded by mountains and there were so many people that night that I couldn't believe it, hundreds and hundreds of people dancing and carrying on. I didn't play one place out of eighty-some places that I played that it wasn't crowded. Always a crowd, just like the early days in America, back in the '20s.

"I played at a big coliseum in Verona and it was the most frightening experience, yet it was marvelous. This is a place the Romans built. The seating capacity was one hundred thousand and they put the opera on there. I was there with a quartet of rhythm and me alone out on that huge stage and I couldn't see the audience because of the floodlight in my face. After the first number the applause was so frightening I almost fell over backwards. I couldn't tell how many people were out there, but there were thousands. I had to do an encore after I had done my three numbers that I was allotted to do and I still didn't get to see the audience because I went backstage where the opera dressing rooms are and where they used to keep the lions and things, and the place cleared out before I got to see what the coliseum looked like. It was quite an experience. They love jazz, that's what I'm trying to say."

Earl Hines assured me, "The foreign people are a little different than the people in the U.S. They dig up these books and magazines and biographies and what-have-you of artists that they like and they know an awful lot about you before you come over. They make it their hobby to study about the artists and they really realize what the artist means when he comes over there. Here, there are any number of people that don't know, or some of them say, 'Oh, I heard of him,' but they don't pass it on to the youngsters. And some of them are not too much interested in jazz. But over there they're interested in jazz from the child up and they know all about you. It makes no difference where you go—and I've gone all over Europe, all the Scandinavian countries, Russia, also Japan, Australia, South America—it's one language with them, see—jazz. They appreciate it so much that when you have a concert there that starts at eight o'clock, if you're not there at eight o'clock, the doors are shut and you can't get in until the intermission."

Mercer Ellington purchased a home in Copenhagen three years before I interviewed him, and the conversation soon turned to Europe. Mercer had had in his custody for more than a decade a huge collection, an archive really, of tapes of performances of his father's band, tapes of interviews with Duke, musical scores, and memorabilia. Since no offer to house them had materialized on this side of the ocean, Mercer accepted one from Danish Radio to become custodians of the collection. The story has a much happier ending than that account implies, for the Smithsonian Institution became, in 1988, the final resting place for this source of inestimable value. However, Mercer's thoughts on the matter, as conveyed to me, provide yet another commentary on the disturbing pattern of neglect of the art form of jazz in its place of origin.

"For some ten years I had these tapes, all the while risking that they dry out or suffer some sort of damage. In desperation, after trying to do something to restore them—and there are forty-one boxes of them—I got through one box at the cost of $5000, transferring them and treating them with Dolby and all the things that make for good sound and so forth. I was told that the folks at Danish Radio would be very glad to transfer them, preserve them, restore them, become the custodians, and use the tapes as a source. I basically didn't give them the tapes. They have just taken the great trouble to become the custodians of the tapes. It's not just music. There are many interviews and things of that sort which we collected. A lot of times Pops, after he had an interview, he'd ask the folks to send him a copy.

"So they have these tapes and from time to time they have a broadcast, every Sunday, of Duke Ellington, taken from these tapes, on Danish Radio, augmented by some of the interviews. So there is a mindfulness kept of him, and this is one of the things that I asked. And as far as the memorabilia is concerned, we were looking for places to put that, but there was no one who would guarantee us that they wouldn't be stacked in some dry room to collect dust, rather than be out where people could know something about him one day. As a result, once they got it, they made a worldwide press release about it and that's what caused the attention. I had considered various possibilities and the main thing is we had to be involved with somebody who really cares, and they do.

"Jazz is American culture and blacks have contributed heavily to it," Mercer continued, "and the reason I feel it's very important is because it *is* what is the United States. If you were in Denmark you could easily feel the strength and influence that the U.S. has on the rest of the world. For instance, they have an institute in Copenhagen which is dedicated to the commemoration of Ben Webster, and Ben Webster lived there for only four or five years. They are so highly in mind of this man that once a year they have a three-day affair which is of a beneficial nature. The funds are contributed to some society or organization or cause. While in the U.S. you can travel from one corner of St. Joseph, Missouri, to the other and you will not see Coleman Hawkins' name anywhere. This is one of those things where maybe the people are too close to the forest to see the trees. When it's your own you don't place the value on it that you do when it's in somebody else's backyard. One of the things that we particularly noticed when we did a tour of Russia for five weeks is the great development and the back-up that they give their culture, their arts. I think we're much richer in what we can produce and do produce. I think America has got a lot to prove."

The bluesman Johnny Shines, who lent us much insight into the idiom of which he is a master in the preceding chapter, was eloquent on this same theme: "This is something that I really don't like to talk about because it's a bad reflection on you and I. Being an American, black American, I'm very enthusiastic about the American people as a whole. But the European people is more appreciative to the type of music that I'm doing than the American people are, and I feel like that it seems the young mother that birthed the baby has leave it on somebody else's doorstep. The blues was born here in America, America's the mother of the blues. Why take it somewhere else? Why don't we learn to appreciate our own, why don't we look at the grass in our own backyard? It's just as green as it is anywhere else. Why do we take it as a past event, not a present thing? It is from the past, the present, and it will be the future.

"The blues will never die because it is the American heritage. The blues was born here in America. The black people being brought from Africa and other parts of the black continent, we brought nothing here with us. Everything that we have attained, obtained, is from here in America. The blues was born here in America out of the bowels of black people, and the blues will forever live because it *is* our heritage, we can't forget it. Not only as a black American, but as an American as a whole, the blues will live in our souls and will never die."

Not to ignore the impact that jazz has made in other parts of the world, we have assembled some brief remarks on the reception of the idiom in Japan.

"I've been told I'm more well known in Japan than I am here," pianist John Coates, Jr., told me. "In Japan, more than anywhere else—it's incredible—they know in what years, in what *months*, you did such and such a thing." John related that, several years before our conversation, the Japanese monthly *Swing Journal*, a lavish publication devoted to jazz, sent a writer to interview him at the Deerhead Inn in

Delaware Water Gap, Pennsylvania, where he had been in residence as house pianist for more than two decades. Subsequently a seven-page article was published in the magazine, his nine albums on an American label were reissued by a Japanese record company, his latest album was cited as the Gold Disc of the Month in Japan, and he was invited to do several tours of the country.

Singer Carol Sloane was making annual tours of Japan in the 1980s and she released seven albums there in the first three or four years of that decade. "They want to hear those old standard tunes," she explained, "like 'Misty' and 'Stardust.' I have never seen *any* kind of concert not filled in that country—rock 'n' roll, the Juilliard String Quartet, country music, *anything*. Possibly the reason that country is so damn successful in what it does is because they're so open to everything."

In December of 1986 I traveled as a journalist with two jazz singers, Deater O'Neill and K. Shalong, their support combos, and the D.C. Contemporary Dance Company to Bangkok, Thailand. The three performing groups represented the Washington, D.C., half of a cultural exchange. I had the opportunity during the intermission of one of the concerts the three D.C. groups presented to interview two young Thai natives who were intimately involved with jazz in Bangkok. They were Pat Sangthum, manager of the Thailand branch of the international division of Polygram/Polydor, and Martina Jalleh, a deejay on the city's first English-language FM frequency.

Martina's show aired four hours a day out of twenty-one hours—5 a.m. to 2 a.m.—of jazz and pop format. Her programming style blended contemporary artists like Grover Washington, Jr. and artists of the past like Duke Ellington and Miles Davis. Conceding that "the people over here are not very quick to accept jazz," she nevertheless estimated the city's jazz audience as "at least five thousand." Letters come to her at the station, she pointed out, that indicate "the people who love jazz, they really love it."

"I'm a jazz fanatic," Pat Sangthum confessed. "I listen to jazz and I know most of the numbers. I was born here, I grew up here, but I started listening to jazz when I bought this record of Sergio Mendes and then I developed from there. You know, it started from Brazilian jazz through the '40's, then back to the '30s, and I'm into vocal jazz.

"Concerts indicate the popularity of jazz here," he continued. "Like Sadao Watanabe, who is getting a lot more popularity in the U.S.A., performed here twice and both of his concerts were sold out. And Lee Ritenour, Sergio Mendes, other jazz figures also came to perform and their concerts were quite successful. So I would say that we have a great number of jazz listeners. I work with Polygram, and our marketing research sort of indicates the popularity of jazz music here. We import jazz records—all the labels—and they are sold out all the time.

It's kind of low-profile here but I would say that there are more peo-ple than you would ever thought that listen to jazz here."

We cannot leave the subject of jazz in Bangkok without mentioning that the nation's first family is headed by an accomplished jazz clari-netist and saxophonist. King Phumiphol Aduldet, who has reigned since 1946, holds a traditional-style jam session weekly, I was in-formed by Manrat Srikaranonda, a pianist who has participated in the Friday evening get-togethers for better than three decades. Over the years the king has numbered among his friends the likes of Benny Goodman and Jack Teagarden, who sat in with this hip royal cat whenever they visited Bangkok.

Unfortunately our visit to the city coincided with a week-long birth-day celebration for the king, and his official schedule was too full to allow him time to make the scene with our singers and musicians. Too bad, really, for we were all anxious to hear him blow.

Manrat did put together a late-night party for us at the Lord Cham-berlain's villa after one of our concerts and it was rumored that the king would drop by. He didn't show, but Manrat occupied the piano bench for a while and acquitted himself well, which wasn't surprising in view of this bank vice president's part-time duties as professor of jazz arranging at the local university, a role that he was anticipating to assume full-time upon retiring in a couple of years. Oh, I forgot to mention that Manrat had attended Boston's Berklee College of Music in the late 1960s. One of his fellow students was the vibraphone player Gary Burton. All of which just goes to prove that in jazz, as in other areas of life, it is indeed a small world.

So far in this chapter we have dealt with jazz's reception overseas. We would now like to let some of the players from abroad relate how they first came into contact with jazz and were inspired to become jazz musicians. In one case we shall hear the story of one who became, not a musician, but an impresario on a rather grand scale, and in another account of one who became both musician and impresario.

"I went to two blind schools and through both of the blind schools studied with the same teacher, a blind teacher," George Shearing, the English-born pianist, began an interview with me. "Between the ages of four and twelve I had a kind of a general music education. There wasn't anything specialized, they concentrated on me between the ages of twelve and sixteen when I took a little bit of theory, not harmony courses and counterpoint, just rudiments of music, as it were, so that I would have an idea the way sighted music was written, and I studied braille music.

"My teacher told my parents when I was sixteen and ready to leave school, he said, 'It would be a waste of time for this boy to study classical music any further. It's obvious that he's going to become a fine jazz pianist.' And so I gave up classical music until I was twenty-one. In the meantime, I started in an English pub and I was playing kind of show tunes and that kind of thing, the more corny show-type tunes. I stayed there for about a year and from there I went with a semi-professional band and I was starting to get an ear for stock arrangements that are written for bands. Then I went with an all-blind band, fifteen blind musicians taught to play instruments from being basket makers, chair caners and so on, and we were playing arrangements by Lunceford, Ellington arrangements. I remember we played Lunceford's 'Stratosphere' and Ellington's 'Caravan' and all kinds of things like that.

"So now I was really getting into jazz and some of the guys used to go out and buy the latest 78 records and bring them back to the hotel room. 'Boy, I bought a real *sender*,' you know, those old terms," George chuckled at the memory, "they used to use back in the '30s. 'This one is a real sender!' It was Fletcher Henderson's so-and-so, Art Tatum's so-and-so.

"Then, through the drummer of that band, I met Leonard Feather and this was about 1937. Leonard got me my first recording session, he got me some dates with the BBC. And of course the war started and the BBC started to call me quite frequently because most of the guys were called up and I had the business pretty much to myself during the war. I would be in the studio in the daytime, theater in the evening, and then a club or a restaurant after that. So I'd be holding down about three jobs on a steady basis. And I used to orchestrate for Ted Heath, I was on tour with a contingent of the [Bert] Ambrose band, I toured with Stephane Grappelli. In short, I was a kind of a high-priced sideman over there and I was a soloist, too, and I had my own band just to play in a club on Sunday nights, oh, sometimes Dixieland, sometimes John Kirby type of thing. That was very enjoyable.

"I came over here first in '46 for a three-month vacation, went back and sold my home and came back on immigration in '47, formed the quintet in '49. In the interim I played intermission piano opposite people like Sarah Vaughan and Ella Fitzgerald and then formed a quartet, which was to be known as the Buddy DeFranco–George Shearing Quartet or the George Shearing–Buddy DeFranco Quartet, depending on which of our wives felt stronger that day."

Another whose experiences with jazz go back to pre-WW II times is the Dutchman Paul Acket, whose annual North Sea Jazz Festival seems destined to go down in history as the Ringling Brothers and Barnum & Bailey of jazz gatherings. There is no question that it has about it a circus atmosphere and, as far as jazz goes, it is without question "The Greatest Show on Earth." With its one thousand or so performers spread across a three-day weekend for fifty acts a day on thirteen or fourteen stages operating simultaneously, Paul Acket's bash dwarfs all competitors bar none. With total ticket sales surpassing

75,000, the mammoth Congress Center of The Hague is jam-packed with as many as 25,000 bodies at any given time during 4 p.m. and 2 a.m. on any one of the four evenings. Indeed, it can get a little wild.

I have attended the festival several times and on one of those visits I asked this affable Dutchman Paul Acket to sit down with me for an interview. It was not until around midnight of the third and final evening of the 1988 event that I managed to chase the man down. He told me that he had first heard jazz on recordings of Ellington, Lunceford, and Fletcher Henderson about 1936.

"Then right before the war—I guess it was in 1939—I started organizing jazz concerts when I was still in school," Paul commenced. "Then during the German occupation—I was living in Hilversum, the radio city of Holland, television you didn't have yet—I organized jazz concerts in normal theaters, not hidden. There were a number of jazz groups from the surroundings and also from other parts of Holland who performed there, like Rita Reys.

"But British and American music was forbidden by that time and what was dangerous was that they were singing in English," Paul laughed. "Nobody took notice and for about one year and a half nothing happened. Now the hall holded about a thousand seats and there were always at least one hundred German soldiers who clapped and stamped their feet. They didn't hear that for a long time in Germany. After about one year and a half I had to go to the highest representative of the Germans in Hilversum.

"He said, 'Kid, you're just a kid now and therefore I won't do anything, but if you do it once more we pick you up and you go to jail.' I said, 'Okay, okay,'" Paul laughed again, "and I went home near Hilversum, a little village called Bussum. After two shows there it was over and then I did it in—how do you call it?—hidden places, underground places. I did that until 1943 and then it was impossible and too dangerous to go on.

"After the war I started again in 1945. Hilversum was the leave center for Dutch and we were liberated by the Canadians, mainly, and some Polish, some Americans, and I did jazz concerts in a very little place holding only twelve hundred seats. It was just the beginning after the war. I was good friends with Charles Delaunay from France and he was the first impresario who brought over Americans to Europe and from him I got Dizzy Gillespie and Sidney Bechet and other attractions. And that started my career in the jazz business with Americans.

"I started the North Sea Jazz Festival in 1976. I always call it a social happening. You must not be mistaken, Holland is not really a jazz-loving country. In Belgium and Germany, France, interest is much more than in Holland. I saw many, many jazz festivals in the past and I was always thinking, we must have it in Holland. But I was sure that if I will do it like in other countries, it would be a complete failure. So I had the idea, also because of this building, the Congress Center, to use five stages, to have the concerts simultaneously.

"So I started with five stages and the first year the attendance was very poor. We had only nine thousand people and a very big loss. We had very good attractions like Ray Charles, Ella Fitzgerald, Count Basie. I thought, well, this proves it's impossible, but the reactions in the press and from the people were overwhelming and then I thought, let's try it once more, hoping not to have the same loss, and the next year we almost double, seventeen thousand, and then I went on and on.

"In some ways it's a hobby of mine," Paul finished up our conversation. "I'm starting organizing it in September and October, alone with my wife, who's in the business, who's working hard on the festival, and from January on I have a secretary, in February another person is coming in, ending up with about twelve people in the office. It's too small, the number of people who are working on it. The last weeks they were working day and night."

Paul has two daughters who are interested in jazz and that circumstance and his parting remark led me to believe that the hosting of his annual weekend jazz party, his "social happening," may be passed down the generations in the Acket family.

"I was in a convent when I first heard jazz," German-born, Washington, D.C.–based vibraphonist Lennie Cujé revealed to me. "Yeah, I was hiding there to sit out the rest of the war. I was twelve and had been attending a music school in Frankfurt. We were bombed out and evacuated down to the Danube near Ulm. That's where the First French Army came through and we were pulled in at the last minute by the retreating German army there and were trained. I became a machine gunner and my ammo 'feeder' was ten years old. We were taken prisoner by the French troops. The little boy stuck with me and we escaped from the open prison camp. We jumped in the Danube and hid out in this convent.

"We were hiding there for about two months. I'd never heard jazz and while we were there I heard this strange music over the Armed Forces Network," said classically trained (on trumpet) Lennie, whose father conducted the Frankfurt symphony before the war. "At first the French nuns told me it was African music, but then when I got to the American sector, trying to find my mother and all, they had the same music playing. I said, 'Oh, African music,' and they said, 'No, American.' So that was it for me."

Lennie didn't hesitate a moment when I asked him who his main influence was. "Dizzy Gillespie—that's what made me come over here. I heard his music and I knew I had to go to the States." He came to the U.S. in the late 1940s and while playing for a high school dance his front teeth were knocked out by "someone tapping on my trumpet."

After serving in the Korean War Lennie settled in Washington, D.C., and began the study of the vibraphone. For most of the '60s he was active in the New York scene with Paul Bley, Larry Coryell, Jimmy Garrison, Albert Dailey, and many others. Off and on for four decades Lennie Cujé has been an important figure on the D.C. jazz scene, playing with tenorist Buck Hill, pianist Elsworth Gibson, bassists Wilbur Little, Walter Booker, Charles Ables, and Steve Novosel, and even the U.S. Navy Commodores as guest soloist.

Another whose introduction to jazz came about through the war was François Pessa, pianist of the Old School Band, one of a dozen or so traditional-style jazz bands in Geneva, Switzerland. We talked when the band visited Washington. "During the war we heard a lot of broadcasting for the Americans and some G.I.s brought some jazz records with them and just after the war a lot of people coming to Europe brought records and I think it all started with that. It was something different and not some music that people could not understand.

"In Geneva we are more close to the black music, New Orleans music," François explained, citing King Oliver, Jelly Roll Morton, and Louis Armstrong as prime sources, as well as jazz artists visiting Geneva, including trumpeter Bill Coleman, a longtime expatriate to France, the New Orleans clarinetist Albert Nicholas, and trumpeter Doc Cheatham, all of whom had appeared as guest performers with the band.

"We have a lot of people who just take it like a kind of religion," François observed, "but in our band we always thought it was a music of fun and we play the music that the people love for the guy who works all day and comes to listen in the evening and be comfortable." As to what motivates him and the other members of the band to devote their evenings to this music that came into being across the ocean before the turn of the century, François was adamant: "We *need* to play!"

At around the same time that German Lennie Cujé and Swiss François Pessa were being introduced to jazz through Armed Forces network programming and records carried in the barracks bags of G.I.s and the suitcases of postwar American tourists, a teen-aged black native South African, Adolph Johannes Brand, was initiated into the clan. He would later be dubbed Dollar Brand by American sailors who frequented the Capetown bars in which this brilliant pianist performed during the 1950s. In the mid-'70s he would convert to Islam and thereafter be known as Abdullah Ibrahim. Abdullah left his homeland in 1962 and resided in Europe. In 1977 he settled in New York.

"The African Methodist Episcopal churches are a very, very strong influence," Abdullah was careful to point out to me. "My grandmother was a founding member of the AME Church there, and there is a very strong following. . . . Richard Allen created the church when he was refused entry into a white church in Philadelphia, and the mother church is there.

"I mean, it is a very, very large community in South Africa and that's how I was exposed to the spirituals, African-American spirituals, and then, through records, to the music, which at that time was like Basie, Louis Jordan, Erskine Hawkins, and the boogie-woogie pianists Albert Ammons, Meade Lux Lewis, and Pete Johnson. When I first heard those boogie-woogie pianists it was an immediate turn-on. I remember we used to play competitions in Capetown, you know, play boogie-woogie non-stop one hour, and that was laying down the basis.

"When we listened to Duke, he was never an American. Duke to us was just the grand old man of the village, of the community. In fact, all the American musicians, even later with Parker, with Monk, with Dizzy, with all the American musicians, we never really thought of them as Americans. They were just the extended family, and that's what it is anyway. The language has changed but the voice is the same."

It was Duke Ellington who, upon hearing Abdullah play in Zurich not long after the South African had moved there, arranged a recording session for him, and several years later Abdullah substituted for Duke on a tour of the Ellington band. "I think it was the highest honor that can be afforded to anyone," he told me. "Listening to Ellington on record is something, hearing the band live was another experience, but being inside it when it was playing, it was like getting on a spaceship, getting involved in the midst of its sound, because it was really at a very mystical level.

"Because Ellington was the ultimate sound scientist. With Strayhorn he knew exactly the combination of instruments to use through experimentation, which colors to use in order to convey a certain feeling. So when I sat inside the band I couldn't play, it was like going into meditation or going into church. It was such a moving experience, there was hardly time to play. As Duke said, of course the most important part of music was listening to it."

Abdullah met Thelonious Monk in Zurich in 1964. "When I started playing piano, started getting involved in my own expression, people always used to say, 'Ah, that sounds like Monk. And so when I finally heard Monk, I said, 'Ah, *this* is the direction.' Monk's music sounded so natural to me. The easiest way to check out Monk is to play him for kids. And they immediately respond to him.

"When I finally met Monk, I said, 'Thank you for this inspiration.' He said, 'You know, you are the first piano player to tell me that.'

"We have a piece that's called 'For Monk.' The first movement I wrote—at least it came—in the '60s. And the second movement—we were playing in Geneva in 1982 and we had a sound check and this piece started coming. It had started like two or three days before, and about three o'clock in the afternoon the whole piece came. Someone said, 'That sounds like Monk.' That evening we received a call to say that Monk had just passed into a coma.

"We can never really think about 'I.' Even if it's solo it's 'We.' *We* are the community. *I* am not a composer by myself, it comes through the creator of the community, and between me and the community, it's *us*."

As the interview drew to a close, I could not resist revealing to Abdullah that my own introduction to the music, in early 1943, had also been via those three masters of boogie-woogie whom he had named as early influences and that to this day those rumbling ostinato piano figures are the sounds that reach deepest into my soul.

"Welcome to the club," he responded.

We continue with the theme of boogie-woogie. "The only problem I had with the piano was that we didn't own one," British bluesman John Mayall insisted. He had already clarified that those very same boogie-woogie giants (plus Cripple Clarence Lofton, another mutual favorite) had not much later turned him inside out with the absolute need to master their idiom. "I had to learn on other people's pianos," John said, recalling the mid-1940s in his native Manchester where he began teaching himself to play boogie-woogie by ear in the parlors of "anybody, really, in the neighborhood who had a piano. It was quite painful for them so they just kept out of the way, shut the door on me and went about their business. It took me perhaps a year before I could get an independent left hand going."

John's jazz guitarist father taught him to read chord symbols but he has never learned to read or write music. "It's totally beyond my comprehension to this day," he confessed. This from an artist who has mastered keyboards, guitar, harmonica, singing, composing, and arranging. A score or more of English blues players has been schooled in John Mayall's Bluesbreakers since the early 1960s, including Eric Clapton and Mick Fleetwood, and this Father of the British Blues, who moved to Los Angeles in 1968, has also worked with such jazz musicians as Blue Mitchell and Red Holloway.

"It's a universal means of expression," John added, "a feeling you get from listening to it and it just goes on and on—the blues will always endure."

There have been in the history of the music some few jazz musicians who have made significant contributions as impresarios, perhaps most notably the guitarist, nightclub owner, radio and television emcee, and journalist Eddie Condon, and festival producer and pianist George Wein. In the mid-1980s I met and became a friend of another such wearer of several hats: the Swiss pianist, bandleader, composer, arranger, and jazz festival artistic director George Gruntz. From several interviews and correspondence with George I have put together his career history. Here at the outset I should note that the occasion of our initial meeting was the 1986 Jazzfest Berlin. That year marked the fifteenth season that he had directed the event.

George, who was born in Basel, Switzerland, in 1932, told me, "My whole youth was carried by music. I'd never been to a football game until I was thirty years old! I grew up in a very musical family, all spare-time musicians, but with great devotion and concentration on 'good' music. I was my father's accompanist on piano since the earliest childhood age, with him playing violin and singing with a beautiful tenor voice, playing the whole popular European classical repertoire, as well as going into cabaret and light music, usually towards the end of a program. I played sometimes for my mother only and I played in family meetings where a whole orchestra was created from more family members. I also played at performances in musical circles of which my father was a member of many of them.

"At the age of about fourteen my father took me downtown whenever jazz was to be heard, as he did to performances of all kinds of music. I got touched deeply by the jazz I heard then, mainly when it was bebop. My first record I bought was Dizzy's 'Anthropology.'

"For some time I followed classical training in piano, later in theory, parallel and as a thing of its own next to my autodidactical endeavors in jazz. At about fifteen I joined a jazz band as accordionist—I wouldn't dare use the piano to 'jazz up'—later playing the vibraphone. This was in the Catholic youth movement, of which I was a member. All the jazz cats in that band were over twenty years old. Soon after getting in that band I was fired from my Catholic Boy Scouts squad for not wanting to leave the jazz band, where 'evil is, and you're far too young.'

"Raised as a 'classical European musician' at home—although with a large view over music in general—I became a jazz musician not only because of my own personal revolution but by seeing the promise that this is music where one is to invent every day and not [have] to repeat what other people have conceived.

"Anyway, the parallelism of me playing classical piano, at Muais Academy, and mainly vibes, for jazz, went on throughout my twentieth year. I became a professional after quitting the Swiss army soon after my twentieth birthday, at that time interrupting another parallel training as a 'high voltage engineer.' My father not only taught me music but the old truth to get trained as much as possible so that one could enter a 'normal' profession once jazz and/or music wouldn't work. I didn't graduate as an engineer but loved this direction and got some degrees as draftsman and construction designer. I never worked in this profession.

"I became first known in jazz circles as a vibist, an instrument that I can't play anymore, doing my first two professional jazz years next to, at that time, great European jazz pianists, like Francis Copieters from Belgium, from who I was still learning. Although I got work around Switzerland, I left with my vibraharp for Sweden, at that time known in Switzerland as the Promised Land of Jazz. I got to jam there with the, at that time, greats such as Lars Gullin, Arne Domnerus, and Simon Brehm, but due to union regulations couldn't work. After tries in Amsterdam and Paris I went back to Switzerland rather disillusioned about the professional possibilities.

"I was looking for a job that would allow me free time to play whenever I wanted and got a job as a car salesman, thanks to a music-loving boss. I didn't want to drive cabs and also didn't want to enter one of the many European radio big bands that never played this type of creative jazz I was interested in. These years of artistic freedom were very important. I only had to accept a gig that I really wanted.

"I switched completely to piano and started to play international jazz festivals, the first one with Switzerland's foremost bebop altoist and jazz pioneer Flavio Ambrosetti, father of Franco Ambrosetti, in San Remo, Italy. That was soon followed by the invitation to play Newport, Rhode Island, with Marshall Brown's Newport International Youth Band in 1958. Coming home from Newport, my life changed. Now I was getting gigs all over Europe, and American artists like Art Farmer, Johnny Griffin, and Dexter Gordon asked me to rhythm-section for their European tours. In 1962 vocalist Helen Merrill took me to Japan for a couple of months as her pianist, bandleader, and arranger, which made me, shortly before my thirtieth birthday, definitely walk away from that one-and-only car salesman job that had helped me a lot to keep my musical direction throughout these years. My boss made me a farewell gift of his own beautiful Bechstein grand, saying something like, 'You're leaving me the way I knew it to happen one day. Don't forget me. Take this instrument that you played so nicely when visiting during those years of cooperation.' Touchy shit!

"Throughout the '60s I was rhythm-sectioning all over Europe and, together with Swiss drummer Daniel Humair, had the most sought-after trio for comping, running around Europe now comping everybody ranging from Roland Kirk to Gerry Mulligan. This trio became the famous Rhythm Machine with Phil Woods in 1968.

"Again parallel, I started more and more to compose, doing, for instance, in 1967, my album *Noon in Tunisia,* jazz and the music of the Tunisian Bedouins, that I also took on tour, mainly with Jean-Luc Ponty, and later Don Cherry, on the jazz side. After one-half year with Phil Woods I left the Rhythm Machine, now, at the beginning of the '70s, going more and more into composition and directing, at first for cult movie soundtracks, later for drama theater, in 1971 accepting an offer as musical director, then chief musical director, at Zurich's famous State Theatre.

"My producing a few special jazz concerts at the State Theatre made me get in 1972 an offer from Berlin to become its program director/ producer. It became the longest job in one place in my life! I really wanted to do it only for a couple of years, two or three, and wanted, like always, just to learn a little about this big festival, the only real jazz festival in Europe then. I was able to learn a lot, maybe more than I wanted. The main reason for me to last the first eight years was that from year to year I fought for the better and the constant promise kept me doing it. Only after 1980, when the city took over the organization of the fest, I started to be lucky with the job—now another reason for *not* quitting!

"I think I have invented this system that has become the one of many other jazz festivals in Europe. Out of the nowadays very big market of *quality jazz,* ranging from almost pop to almost European classical, one can only produce a serious jazz festival by emphasizing a few themes that, in my case, are representative of the actual scenes. Now I never did use themes like one-evening pots—a Trombone Night, a Bebop Night. I carried the themes like tracks throughout the program. So in an evening program with three groups one would be trombone, one bebop, one who knows, with usually four themes and three groups a night. The program thus becomes colorful, undidactic, and inviting for people that come for one theme and enjoy two more. I personally like to do it that way and it covers my own concept of work as a jazz artist who believes in the promise of jazz to be something wide open with great models of tolerance rather than the incest of just one little edge."

When I asked George to expatiate upon the theme of the role jazz has played in European culture, he displayed true insight and offered a sort of synthesis of so much that I had been given in bits and pieces from other sources.

"Jazz was never really 'entertainment' in Europe," he began, "it was always respected as something special. At first, at least, it was not as easy to understand and even more difficult to play, always something 'cultural.' That's why jazz in Europe often was boring. Exceptions like Django Reinhardt prove my word. Still today an impresario like George Wein says, 'It's difficult to produce European jazz artists on my festival because they don't present themselves. They play very well but they look like shit.' That sounds hard but there is some truth there. Only recently the 'serious fun' element has even become an European invention, not only in my work, but mainly in that of, for instance, the Vienna Art Orchestra, Mike Westbrook, et cetera.

"Jazz became faster than anywhere in the world here in Europe a 'cultural thing' subsidized by cities, countries, and cultural organizations. Benny Goodman's Carnegie Hall concert was a big thing in the U.S.A. It wouldn't be here, at least in terms of taking place in a holy classical temple. Here in Europe all Philharmonic Halls were always open to jazz, and the connected budgets were also!

"There is, of course, the social aspect of many—too many?—jazz artists living in the U.S. and overloading the market at home. New York is terribly full of talent that has not enough places to be presented. So this music and its musicians become 'cheap' on the market. Whereas in Europe they are visiting guests. And often it has happened, if one of these artists thought to remain in Europe where he felt to be respected, after staying two or three years he became local."

Confessing that the name George Gruntz was not especially familiar to this observer of the jazz scene until I attended Jazzfest Berlin, I asked George to share his thoughts with me on whether or not he found this circumstance to be the usual state of affairs.

"When being in the States I am amazed how little jazz people know about me. This is a phenomenon and I don't dare analyze it, but it becomes pathetically rummy when, as on my U.S. tour last October with my Concert Jazz Band, jazz fans incessantly ask, 'Who the hell are you?' Well, that wouldn't stem my love for the U.S., which started to grow out of jazz during my teens and which didn't change much, unlike [with] other Europeans, after Vietnam and whatever. Although European jazz artists have a right for their own proudness that has developed a lot these years parallel to their own stylistic developments.

"Two things are important to me. Jazz, the most important musical invention of this century, was born in the U.S., where musicianship—the technical side—is in best balance with creativity—the inventive side. For many of my projects I can invite European artists to work with me for best results, even better than with U.S. artists! But when it comes to the concept of my Concert Jazz Band, a full band with French horns, my U.S. partners are the best. The band always consists of two-thirds U.S. members, although we then play [a] very European-oriented book!"

Czechoslovakia-born pianist Adam Makowicz, his wife Irena, and I sat talking over a sidewalk cafe table at Washington, D.C.'s One Step Down two years after his arrival in the U.S. in 1977. "My parents were musicians," Adam explained. "My father played mandolin and guitar and my mother was a pianist and singer and a piano teacher. And from when I was very, very young I remember that my parents arranged some kinds of concerts in a private home and usually came a lot of musicians. My parents liked very much Chopin, Mozart, and Beethoven and my father liked to play popular songs." Adam and his family moved to Kraków, Poland, when he was six. "I started classical music when I was nine years old. My parents never said to me, 'You must practice.' It was a pleasure, practicing.

"The first time I heard jazz it was big band—I don't know who it was—in 1956, I think. And when I heard it first time, I knew it was some kind of song for me and I liked it and I knew that I must start to play it, to learn it. I was listening every night then, especially Willis Conover"—that is, the veteran Voice of America jazz broadcaster— "who was best teacher, and still is, for a lot of musicians in east countries of Europe. I just listened because I loved this kind of music, but I didn't know nothing about swing, about improvisation. I learned later about that, the details. Nobody taught me because no teacher in Poland knew about jazz at that time and I must by myself listen and try to play like this what I heard in jazz.

"When I first heard Art Tatum it was really a shock for me. I loved this guy and I loved his music. He was among the first I heard. I listened every night, to Earl Hines, Erroll Garner, Oscar Peterson's trio, Teddy Wilson with Benny Goodman and Gene Krupa. I tried to listen to *everything*! It was a time when it was difficult to find records in Poland. I was sitting at the radio all night."

Irena filled in some of the details. "It was hard time, everything was destroyed, bombed. After few years they started to think about pleasure and song and music, not just right after the war. When Adam started to hear this music he became so involved in jazz that he decided to quit classical training because the teachers didn't allow him to play jazz. They and his mother wanted him to be a great classical piano player, not a jazz player. So he quit his school, escaped from home, and more than one year he lived in the jazz club in Kraków, which was the only and very famous jazz club in this town.

"And he devoted himself completely to jazz. He lived there, he was practicing about ten or twelve hours a day, and he played in the evening some concerts or with groups. In 1961 or '62 he arranged a group with his close friend trumpeter Tomasz Stańko. They formed a group named the Jazz Darings. There was a South Poland competition in jazz and they won the first award. And from that time he came to play professionally jazz."

It wasn't long after this that Adam found himself called upon to accompany the likes of Art Farmer, Ben Webster, and the trumpeter Carmell Jones in Poland and throughout Europe. He worked with Polish singer Urszula Dudziak and in 1976 he played with the Duke Ellington band at a concert in Warsaw. Adam and Irena now live in New York.

"Anywhere you go over there you can find jazz fans," Moscow-born trumpeter Valery Ponomarev, who moved to New York in 1973, told me, adding that he had first heard jazz on Conover's VOA show. "Over there I had access to jazz only through tapes and radio. You tell somebody, 'I want this album,' and he might say, 'Oh, I know one guy, his father goes abroad, maybe he can buy you the record there.' Once I got me a brand-new record of Clifford Brown that way. I paid for it like half my monthly salary, but that was all right."

The multi-reed and woodwind player and leader of the Willem Breuker Kollektief told me that he "was fired after one or two years in the conservatory in Amsterdam." Willem had started on the clarinet when he was eleven and had entered the conservatory upon finishing high school. "They told me, 'You don't have any talent for music so it will be very hard for you in future if you continue making music.' That's what they told me. Now they say to everybody, 'He's one of my pupils.' " The boast of Willem's former teachers, specious though it is, is understandable, considering that the Kollektief began to be a star attraction at festivals throughout Europe during the 1980s and has made several very well received tours of the U.S.

"There were certain rules in the house when I grew up," said Danish bassist Niels-Henning Orsted Pedersen, "that we all had to take piano lessons from a very early age. I am the youngest of five children and everybody else was playing around me including one of my big brothers playing trumpet in a big band. Which means that he bought all the Count Basie records, you name it. So what I can say about jazz is that I've heard it for as long as I can remember. I was given the bass by my parents under one condition, that I would pick up classical music first, which meant that I was taking lessons before I had my first instrument." N.H.O.P., as he is known throughout the jazz world, was playing by the age of sixteen with Bud Powell, Dexter Gordon, and other American jazz artists upon their visits to Copenhagen. He later became a member of Oscar Peterson's trio.

In Calcutta, India, a sitar player, Sanjay Mishra, had his interest in jazz sparked in the 1970s by a performance there of American guitarist Charlie Byrd. Sanjay told me that jazz was a major factor in his decision to emigrate to this country. He settled in Washington, D.C., and in the 1980s was a member of a quartet that blended jazz and classical Indian forms. The other members were his fellow Calcuttan Broto Roy, a tabla player, saxophonist Carl Grubbs, who had received instruction from his cousin Naima's husband John Coltrane, and a young Washington-based pianist, John Kordalewski. The last, reflecting on the coming together of the two cultures, observed, "To me, what makes it exciting is that we're learning about them and they're learning about us."

Bobby Enriquez, who grew up in Bacolod city in the mountainous Philippine province of Negros Occidental, assured me, "I didn't see a TV set until I was sixteen. My sister's famous for committing suicide," Bobby joked. "If my mother didn't give her what she wants, she'll take out my father's forty-five caliber—but there's no magazine in it—and hold it to her head, and that's how we got the piano. She took lessons and I sat in her lap. That's how I learned. I was one year old and a half."

At four Bobby was picking up tunes by ear and at six he won a radio station amateur contest. Sneaking out of the house for late-night paying gigs when he was twelve, Bobby dropped out of school when he was fifteen and ran off to Manila, landing a job in the city's premier jazz club with Vestry Rojaz, who is considered one of the best trumpeters to come out of the Far East. That was only the beginning of a three-decade-long odyssey that took this piano-playing Filipino onto the bandstands of clubs in Hong Kong, Bangkok, Singapore, Okinawa, Taipei, Japan, Hawaii, San Francisco, New York, and Europe.

In the late 1980s Bobby was living in Union City, New Jersey. His dream, he told me, is to establish a music education foundation at the University of the Philippines, making it possible for American jazz artists to lecture and perform for young Filipinos. "See, I had to run away," Bobby explained, "and I would like to give my people some chance, put them in the ring. I don't care if they lose. At least they'll have a feeling of being there. And if they win, we have a champion."

It was saxophonist Paquito D'Rivera's father who introduced him to jazz. "My father was playing saxophone in a big band at the Tropicana in Havana," Paquito informed me, "and I had a chance to see all the stars at that time when I was a very little boy." This experience was repeated over and over, for Paquito's father played in the pit orchestras of many of the city's theaters and often took his son along so that the youngster could watch from the wings. Paquito cited as a pivotal experience the presentation to him by his father, at the age of five, of an album of the 1938 Benny Goodman Carnegie Hall Concert. The difficulty of acquiring records in his homeland at that time was also alluded to by Paquito. "I would love to express how important Willis Conover is for jazz musicians in Cuba," he said, adding that Conover's "Music USA" "was copied at every opportunity."

Paquito came to this country under rather extraordinary—in fact, unique for a jazz artist—circumstances. "Always I have a dream to be a jazz musician in New York," he explained. While on tour in 1980 with the Cuban big band Irakere, Paquito sent a suitcase filled with rocks onto the airplane waiting at the Madrid airport, thereby initiating the defection from Cuba that would bring about the realization of his longtime ambition. "Jazz music was not forbidden in Cuba—it wasn't a law—but in some way you knew it was much better you do other kind of thing than this 'imperialist' music."

Maynard Ferguson, who was born in Verdun, near Montreal, Canada, told me, "At the age of fifteen I had my own band working professionally, and I often tell in the schools when they ask me what my schedule was of playing and practicing during my early years, I say that it would be impossible for *them* to do that because their mother and father would be arrested immediately, 'cause at the age of thirteen I was playing from nine 'til three in the morning six nights a week and on the seventh night we started at two o'clock in the afternoon and went 'til two o'clock in the morning.

"I was getting my schooling by that time at home from my parents, who were teachers, and I was also involved in studying at the French Conservatory of Music in Montreal and things like that. So that was more or less my Canadian scene. I enjoyed the thing of being a bandleader at the age of fifteen where the average age of my band at that time was much older than the band I have right now. It was during those years that I met Stan Kenton and Charlie Barnet and Jimmy Dorsey and Boyd Raeburn and all those bands that I later played with in the United States and they all made standing offers if I ever chose to break up my Canadian band and come to the U.S. Ellington made me the same offer and I would have enjoyed playing with the Ellington band, I think. I was a little disappointed that I somehow never played with Duke, but most of the offers came from Ellington after I'd already started my own band.

"When I broke up the Canadian band," Maynard explained, "it was because I felt that I'd gone as far as I could in Canada at that time. Canada is not as far behind this country musically now as it was then. In fact, they're rather hip now, but in those days it was a very stagnant thing and not much attention was being paid in Canada to the kind of music that originated in this country. So I went with a very creative band, the Boyd Raeborn band, which wasn't that well known but did some of the most inventive and creative things and a lot of great musicians were in it. If you listen to his things today they're still far out."

A word-of-mouth report of Marshall Key's saxophonic prowess had traveled the jazz grapevine to the ears of the producer of International Jazz Week in Mainz, Germany, the young Washington, D.C.–based altoist related to me upon his return from a first stay abroad. He had been invited to participate in the program and so he packed up his horn and a duffel bag of clothes, made the gig, and then wandered around the continent for four months.

"I had a ball," he enthused. "People take jazz more seriously over there." Chance encounters with other American jazz players, including saxophonists Archie Shepp and Pony Poindexter, and "hanging out" at jazz clubs put him in touch with the jazz network there. As a result he worked at least two or three jobs a week for his entire stay, which he divided between Germany and the Netherlands with a week's visit to Paris. He even left some sounds behind in an album to which he contributed.

"I checked into this really seedy hotel," he recalled of his arrival in Amsterdam, adding that he sat in that first night at Carnegie Hall, a popular jazz bistro, where he heard that the nearby Alto Club needed— you guessed it—an alto player for three weeks. "So I got the gig," Marshall chuckled, "and the place was packed, I mean, literally. In Germany, I did a gig that paid about two hundred and fifty dollars, and the festival I did paid six hundred dollars. I played one set. But I got seriously homesick. I would like to travel to Europe again and do the festivals, but I want to live here."

Tenor saxophonist Houston Person, a seasoned world traveler, assured me that he runs into no language barriers between his horn and non-English speakers. "No, no barrier whatsoever," he insisted. "I guess that's why they call jazz the universal language." On the relative merits of stateside and overseas audiences he had equally firm opinions. "A lot of guys say the audiences over there are greater, but I don't see any difference. An enthusiastic audience is an enthusiastic audience, wherever you find them. And I just love to find them!"

10

The Contemporary Scene

*I think it's maybe the most important move-
ment in jazz in the next decade, the restoration of early jazz.*
                          Bob Wilber

The word fusion, *like the word* jazz, *like
almost any of the sort of standard buzz words
that exist in improvised music, you can't find
any two people to agree on what it is anyway.*
                          Pat Metheny

*As a black female jazz trumpet player, I'm
at the bottom of the barrel. But I refuse to
stay there.*
                          Clora Bryant

*The art of it is . . . to always come up with
something different every time you hit that
bandstand.*
                          Philip Harper

All of the main styles and formats of jazz discussed up to this point in
these pages were still, in 1990, being heard live and being recorded.
In some cases the *style* was being re-created while, separately, a long-
established *format* was serving as a vehicle for evolving forms of the
jazz idiom. For example, the big-band format was still being utilized

along the lines of the Swing Era units and the 1940s bebop big bands while at the same time serving the needs of current avant-garde expression. The same could be said of, to cite another prominent example, the bebop combo approach, which some of the pioneers in that form of expression were still playing more or less as they were over four decades ago, while younger musicians of that inclination were creating an approach dubbed "neo-bop" that has embraced "free" elements introduced in the 1960s, '70s, and '80s.

Nor let us omit mention of such a group as the Dirty Dozen Brass Band, a New Orleans type marching band in design and attack, yet thoroughly modern in terms of its use of r&b and the materials of Thelonious Monk, Charlie Parker, Stevie Wonder, and others. The DDBB more than once has found itself on the same bill as a group of unabashedly antiquarian traditionalists and has won many a new fan in an audience that, likely to an individual, came only to hear the group or groups re-creating the older forms.

The re-creation of earlier styles comes in two principal forms, casual and disciplined. As an example, and a very good one, take the Black Eagle Jazz Band of the Boston area, a septet that, although all of its members work full-time day jobs as professionals in fields other than music, plays two or three gigs a week and often appears at festivals here and abroad. The group's leader and cornet player Tony Pringle has described for me the way in which the members learn a number recorded six decades ago by King Oliver, Jelly Roll Morton, or Louis Armstrong, to name only several of the many sources of the twenty-year-old BEJB.

Seven cassette copies of the original are made and each band member listens repeatedly to his copy as he drives to and from his day-gig workplace. Then, with the outlines of the tune fixed in their musical memories, the seven musicians work out a "head" arrangement at rehearsal, using this as the basis for public performance, when they will improvise solos and ensemble passages in the style of the original artists, usually stretching out the tune to twice, sometimes three or four times, the three-minute recorded version by Oliver or one of the others.

The modus operandi described above is, I gather, a common one for combos like the Black Eagle Jazz Band, which must be numbered among the finest bands worldwide presently playing traditional jazz materials. Other such bands from the U.S. and abroad are well represented on the Stomp Off label, which in a few years has compiled a catalogue of more than a hundred releases devoted to documenting bands playing in the older styles of jazz as we approach the centennial of the idiom.

"My main function in jazz right now," soprano and alto saxophonist and clarinetist Bob Wilber affirmed in an interview with me, "is to try to spread the word on all these great giants—Ellington, Goodman, Jelly Roll Morton, Bechet, and so many others.

"I think it's maybe the most important movement in jazz in the next decade, the restoration of early jazz. It's a re-creative thing. ot just imitating old records, it's being creative in the idioms of earlier music. In other words, we play Mozart today and it isn't considered old-fashioned. Why not play Jelly Roll Morton, King Oliver, early Duke Ellington, Benny Goodman, and Charlie Parker?"

In clarification of his conviction that there is also a need for the *evolution* of earlier forms of jazz, Bob added, "I'm a great supporter of the Dirty Dozen Brass Band, which is a continuation," and he also cited David Murray's Big Band as another valid approach to the renewal of the older tradition. "His [Murray's] theme is a tune he wrote called 'Bechet's Bounce.' "

A good case can be made for Bob Wilber as the leading spokesman for and practitioner of the jazz repertory movement, which was launched in the mid-1970s by the New York Jazz Repertory Company and the Smithsonian Jazz Repertory Ensemble, in both of which he was active.

"The concept of jazz repertory is to present different styles of jazz, re-creating earlier performances. It's kind of a new concept and I think it's a valid one," Bob insisted, citing examples of programs by the groups alluded to above as well as his own units. "We've done concerts involving the music of Duke Ellington, Bix Beiderbecke, Jelly Roll Morton, Benny Goodman, King Oliver, and Sidney Bechet.

"I think you bring this music to life on stage," Bob asserted, "and it adds a dimension that you don't always get from listening to old records. There's a certain amount of direct re-creation of what happened on the original records, but there's always leeway for the individual to express himself. Improvisation is a vital part of jazz and without it jazz is dead music. The re-creation thing is difficult to carry off because, if you attempt to re-create exactly, note for note, say, an old record by Ellington, Armstrong, or Beiderbecke, there's no way you're going to equal the original because the original was created on the spot, the solos were improvised. You can't take an improvised solo and copy it and do it as well as the original creator.

"So this is a problem with re-creations. You can, say, in the case of Ellington, where the voicings and tone colors are an integral part of the music, you can use that and get the flavor, but it's much better, to my way of thinking, to allow the soloists to create their own solos, rather than try to copy the old ones.

"There's very encouragingly quite a lot of young musicians coming up who are able to play in the early idioms very convincingly. I would like to see more young players come up in the universities in jazz studies, re-creating performances of the past through all of the great styles, and in that way learning about jazz by playing, say, in the style of King Oliver's band, or Duke Ellington in 1930, and so forth."

That Bob Wilber is, incidentally, uniquely fitted to be a leader in the jazz repertory movement one may infer from the account of his early history included in our chapter on New York.

Bob Wilber emphasized the need for "leeway for the individual to express himself," if the re-creative effort is to be valid as jazz, and he alluded to the "direct re-creation of what happened on the original records." However, he did not go into any detail regarding the labors required to capture with accuracy "what happened" when those sounds were fixed in time in wax or, later, on tape.

Pianist Bob Greene, considered one of the few who has captured with feeling the essence of Jelly Roll Morton's pianistic signature, told me that when he first began listening to Morton's records in the 1940s and trying to re-create his style, "I couldn't find the harmonies on the piano." Interminable—it seemed—listenings opened up Bob Greene's ears and he became a leading interpreter of Morton's works. "Of course I'd sort of love to be sixteen or seventeen today," Bob confessed, "and have the transcripts available." At the same time, he cautioned, even with the piano score, "you would have to have the sound in your ear before you started, and you just have to love it."

The transcripts referred to were done by another leading Morton-style pianist, Professor James Dapogny, whose five-hundred-page *Ferdinand "Jelly Roll" Morton: The Collected Piano Music* (1982) was the very first complete edition of *any* jazz musician's recorded *oeuvre*. In addition to possessing an uncanny talent for improvising at the keyboard in Morton's style, Jim has also led groups that play à la the Red Hot Peppers, Morton's studio recording group. I asked Jim to talk about the difficulties involved in such re-creation.

"There are many who believe that anything that's New Orleans jazz or Chicago type jazz or anything lumped together under the name Dixieland is essentially a simple kind of style to play in," Jim pointed out. "In the classical musical world, which I have my other foot in, you take for granted the idea that if you're going to play Mozart, you don't play it the same way you play Beethoven. In jazz there is, just as in classical music, this question: In what way do we bow to the performance traditions associated with a style when it was originally conceived?

"If you're going to play Morton, you're going to have to play it in such a way that at least approximates the way people of the time played it. And those principles are not that simple to learn. I mean, it's not just a matter of playing the right notes or notes that happen to fit the chord changes. With Morton's music, for example, there are structural and surface features that make it impossible for it to be played in a modern jazz style. The whole business of improvising as an ensemble is a tradition which the Art Ensemble of Chicago and the Dirty Dozen Brass Band and some others are trying to recover, but in general it's far from being a well-known principle in jazz practice these days.

"Morton combined fixed and improvisational elements at different levels, so it's a rather different idea of how composition and improvisation interact than what is common in the jazz of today. In his music there is a fascinating interweaving of fixed and variable elements. It may be the most important thing that we can learn from Morton, something that even the art music tradition could learn from him. I think Morton was the first to learn to do that. Or maybe not *learn* to do it. Maybe he just had whatever synapses you need to have in your brain to be able to do it intuitively."

The chore of transcribing ensemble passages recorded three and four decades in the past is "quite a job," Washington, D.C., trumpet player Allen Houser assured me. Allen was preparing his sextet for a Smithsonian Institution concert of the music of Tadd Dameron and Horace Silver. "It's a repertory presentation and reflects the same notes as the originals, but there are different colorations because we're using some different instrumentation. Of course, the solos are not transcribed, they're whatever we can come up with at the moment."

The Smithsonian Institution has been in the forefront of jazz repertory activity since the early 1970s when critic Martin Williams first came there as director of the jazz program in the Smithsonian's performing arts division. Having featured repertory jazz regularly throughout the several concert series he organized for each season during the 1970s, Martin made a discovery in the early 1980s: namely, Doug Richards, director of jazz studies at Virginia Commonwealth University. Doug had for several years been gearing the university's jazz ensemble for repertory performance, and in 1984 he provided a stunning program of Duke Ellington's music for a Smithsonian Resident Associate concert. Doug has returned with his ensemble almost yearly, at Martin's invitation, to offer repertory programs of Monk, Basie, and Charlie Parker.

"What would the people who are involved with Western European art music do if Mozart's music were available only on records?" Doug asked rhetorically. "There'd be all kinds of very capable people transcribing this music and it would be done right away. There is nothing more important in jazz ensemble than Duke Ellington's work, and up to this point we don't have anything, really. We need to compile a tremendous amount of literature on a far greater scale by a whole lot of competent individuals, and it's going to take a while."

As an indication of the massive effort that will have to be applied to the transcribing of Ellington's recorded *oeuvre*, Doug described in brief his own labors.

"There were things that were omitted, things that needed to be corrected," said Doug of the sketchy scores that constitute the only visual supplement to the records. "The recording and the score were different and it was very tricky to come up with what the original performance really was. Ellington seemed very guarded about his music ever getting out. Nothing was published.

"I just use a reel-to-reel tape deck, a pencil, and my ears," he explained, "and I've gone over a two-bar passage two, three hundred times, and I'm still unsure. I make sure that all of my students have the tapes of the original performances, and we listen many, many times. The idea is not to play [in performance] every tiny nuance of the original recording but to try to understand, as best we can, the essence of the original performance and then give your best for your own individual interpretation."

For the program of Monk's music that the VCU jazz ensemble provided the Smithsonian in 1986, Doug Richards scored some of the composer's piano solos for full orchestra, used recordings separated by almost two decades as the basis for an arrangement of Monk's "Misterioso," and had "sort of like two New Orleans style bands marching down both aisles" for the opening selection of the concert, "Blue Sphere," which Doug described as "a really priceless improvisation, stride-like, very James P. Johnson, which Monk recorded in London." It's little wonder, in view of these and other examples of Doug's originality of concept, that Martin Williams opined, "Doug has developed an absolutely unique and brilliant way of writing for large jazz ensembles," appending the admonition, "Record companies, wake up!"

I asked Martin why it was important to present the music of Monk in the particular manner employed by Doug Richards and his ensemble. "To me, it's because the way that Monk's music was performed by Monk in his lifetime was very dependent on his presence," Martin replied. "He was always orchestrating, re-orchestrating, composing, re-composing. Now, in a sense, every jazz musician is a spontaneous composer, but this goes much, much deeper than that into the question of orchestration. If you don't have Monk, what are you going to do? It seemed to me that you somehow had to write Monk into the pieces as a part of the orchestration, the way he did himself, spontaneously. This is one way of preserving and keeping his music alive as a performance music.

"I think it's very important to do that because he is one of the three major jazz composers. I mean, if you're going to make a list and you get to third place, he's either in the second or the third place. Just as we now seriously think about preserving and performing Ellington by using his orchestrations and, at this stage of the game, even imitating his soloists note for note—we may get beyond that—and do the same for Jelly Roll Morton, then here's a way of doing it for Monk. But again, Monk's music seems to call for a different kind of orchestration than had ever been done for jazz ensemble before and I think Doug is coming through beautifully."

The reader, and listener, who has not been exposed to the sounds of early jazz may well inquire, why bother with all that ancient history? Vince Giordano, tuba and string bass player, leader of the Nighthawks, and member of sundry other re-creative units, summed it up for me: "We're trying to convince that the old stuff is just as good as the new."

Incidentally, in case you're concerned about the future of jazz repertory, harken to the observations of clarinetist Richard Stoltzman, arguably the greatest living virtuoso on his instrument on the classical side at the time of writing and a good enough jazz player to join the Woody Herman band on tour both before and after its leader's death in 1987. I asked Richard if he thought it would be possible to keep jazz alive, in all of its various styles, forms, and periods, over the course of the centuries to come, and would it be feasible to expect improvisation to remain an element in these re-creations of the distant future?

"I don't see why not,". Richard responded. "You see, improvisation is also contained in the music of past centuries, too, so I don't think there's any problem with that at all. What it will depend upon is whether there is an audience for the music, and that is a fairly safe presumption. I mean, we're talking about excellent music in many different forms. As long as the music moves people and takes them to another place and gives them nourishment, which I think the music we're talking about does, then it's going to live."

Of the several aspects of the contemporary jazz scene surveyed in this final chapter, one deserves consideration apart from the various stylistic approaches to the art form current in the 1980s. We refer to the role female instrumentalists were playing in jazz during this period. True, women have been present in the music since its early years as pianists and as singers, but their entry into the ranks of the brass, reed, and rhythm sections was traditionally and routinely met with such bias and hostility that, for the most part, they found themselves relegated to such all-female bands as the International Sweethearts of Rhythm and Ina Ray Hutton's Melodears.

However, the barriers against women which were intact for so long in jazz began in the 1980s to, if not tumble, at least provide some openings through which talented female instrumentalists could pass as they sought to join their male counterparts in the profession. Here are some accounts of and observations by women who have managed to negotiate those openings to the still male-dominated bastion of horn, reed, and rhythm players in jazz.

I had seen an occasional reference in the jazz literature to the trumpeter Clora Bryant and had in my record collection a reissue of an album by her 1950s combo when I saw her perform at the 1987 North Sea Jazz Festival with the Jeannie and Jimmy Cheatham Blues Band, but I was hardly prepared for the power of playing and sensitivity of expression that greeted my ears when she soloed on up-tempo selections and was featured on the ballad "I Can't Get Started With You." In a word, she knocked me out!

Clora Bryant grew up in Denison, Texas, only several miles from the Oklahoma border and directly north of Dallas. She told me that the Baptist church played a major role in her early background and that her oldest brother, upon going into the armed service in 1941, provided her with her first trumpet.

"He left the horn at home," Clora explained, "and that was the year we got a new principal and he brought in a lot of extracurricular activities like the band. I wanted to be in the band and the trumpet was the only thing that was there. My dad didn't have to buy it because we were too poor to buy anything. We just come off the NRA, you know, with Roosevelt. So I decided I was going to have to play the trumpet. Took up the trumpet at thirteen."

With four years of piano lessons behind her Clora was soon sufficiently proficient on the horn to become a member of the high school marching band as well as the school's "swing band." Her teacher was Conrad Johnson of Houston.

"I always liked Harry James 'cause I'd hear him on the radio all the time and I got this record, 'I Had the Craziest Dream,' and when I learned that I knew I was going to play the trumpet the rest of my life."

Babysitting and washing dishes on the weekend provided the funds for trumpet lessons, and before long Clora, along with some of her high school peers, formed a combo to play proms and gigs at roadhouses, " 'cause we had a good band," she assured me.

Attending Prairie View College, Clora was invited to join the institution's all-female Prairie View Coeds, which had come into being because "the guys were in the service." The orchestra played Houston, San Antonio, Austin, Dallas, Fort Worth, and other locales, providing the intermission entertainment when such groups as the Ink Spots and the band of trumpeter Cootie Williams were the top of the bill. In 1944 the Coeds did an East Coast tour, ending up at the Apollo Theatre in New York.

The following year Clora's father decided to move his family to Los Angeles where he had learned there were work opportunities in the city's shipyards. "I got a chance to get to the Central Avenue scene," Clora pointed out, "and to hear the live broadcasts of Dizzy Gillespie and Charlie Parker," who were in residence in late 1945 and early

1946 at Billy Berg's Swing Club in Hollywood, "and I was just amazed!"

"But, you know, it was the strangest thing," Clora recalled. "When we played the Apollo in New York we stayed at the Cecil Hotel and downstairs was Minton's. And we'd go by there. I couldn't go in, I was too young, but I'd stand outside and hear all this music. I didn't know what the hell it was and they'd move me right along, but when I get to L.A. someone says, 'The same thing we heard at the Cecil Hotel,' " the reference being not only to what Gillespie and Parker were playing at Berg's but to what some of the hipper musicians at the Central Avenue venues were offering.

"At Wallace Music City you could go and sit in a booth and play records and you didn't have to buy them if you didn't want to. And that's what I'd do half the day is go and listen to all these sounds, and then buy a couple of them. I'd use some of the money my dad'd give me to eat, I'd go and buy the record and listen at night."

Clora's brother took her over to Central Avenue so that she could transfer to the black musicians' union local. She was seventeen. "I'd have to stand outside at the Down Beat when Howard McGhee and Teddy Edwards and Roy Porter were there."

In 1947 Clora was attending jam sessions and playing in clubs, sometimes in an all-female group in Hollywood, other times in Central Avenue's Club Alabam in a show band with saxophonists Wardell Gray and, later, Frank Morgan, accompanying the acts of such performers as Josephine Baker, Billie Holiday, and Al Hibbler.

"I played every club that was on Central Avenue. I played the jam sessions at the Down Beat. At that time jam sessions were considered just like jobs. There was a whole different conception of the way the music went down. Like when you say you played with someone, nowadays they think you have to refer to being employed, but in those days, if you played in a jam session with people, you were playing with them.

"I didn't run into any obstacles because when I started sitting in on the sessions the guys figured if I had the nerve to get up there—. I was the only female out there who was going to the sessions and actually getting up there playing. [Saxophonist] Vi Redd was kind of sheltered, her father wouldn't let her hang out, and Melba Liston was playing with Gerald Wilson's band at the Alabam and she played with Bardu Ali at the Lincoln Theatre, but she never would go to the jam sessions. I mean, I got up there and challenged Al Killian for high note! The Basket Room, that was one of the leading jam sessions, the Bird in the Basket. Hampton Hawes and all them."

In the early 1950s Clora played the lighthouse Cafe at Hermosa Beach with Howard Rumsey, Teddy Edwards, Shelly Manne, and Hawes. By this time she was a full-time self-supporting jazz musician, playing Las Vegas with the Billy Williams Revue for six-month en-

gagements, on the road for the rest of the year. She recalls meeting the
trumpeter Wynton Marsalis in the early 1960s when she was a guest
in the Marsalis home in New Orleans. "He was crawling around on
the floor," she chuckles at the memory. "I changed his diaper." Clora
gave birth to two sons in the mid-'60s and this curtailed her mobility.
"It wasn't really that slow, it was just that I couldn't really travel and
do what I wanted to do. So I only did what they call 'casuals.' "

In the 1980s it began to pick up for Clora. She studied music at
UCLA, worked with Johnny Otis, who took her to Europe for her
first visit abroad, played the North Sea Jazz Festival with the Cheatham band, and took her own group, with Eric Reed at the piano and
sons Kevin Milton (drums) and Darrin Milton (vocal), to the U.S.S.R.

"The competition is very great," Clora observed, reviewing the difficulties, and absence thereof during some periods of her career, of
playing a horn in the virtually all-male world of the jazz life. "When
Blue Mitchell died I took his chair in the Bill Berry band, but there
was a little—I don't know what the word is, but Bill let me go because
of, you know, dissension. There was resentment. There is a lot of that
and that kept me from getting a lot of jobs. That's why I took the
initiative to do what I did and that's why I *am* taking the initiative.
Before, I'd sit back and wait.

"I know who I am and I know what I can do, but people don't give
you a chance to do it. You got to get out there and do it yourself.
When you put a horn up to your mouth or inside your mouth, like a
sax, it's hard. I mean, you've got all kinds of strikes against you. It's
going to be hard for them," advised Clora, alluding to several young
female horn players whom she knows. "It won't be like when I was
coming along 'cause, see, there were a lot of girls coming along. At
the time I started, most of the men were in the service, so we didn't
have too much of a challenge there. So you could kind of step into
that, and then, if you wanted to continue, the door at least was open,
if you had the nerve and the motivation enough to get out there.
Also, there were all kinds of girls' groups out there. But then, when
the men came back, the scene changed a little bit. But it didn't change
that much for me because I was established in Los Angeles by that
time and I was getting some good gigs."

I related to Clora an observation made to me by trombonist Melba
Liston and she echoed the sentiment. "She's right. That's the same
with me. As a black female jazz *trumpet* player, I'm at the bottom of
the barrel, too. But I refuse to stay there, I refuse to stay there. I'm
not going to stay at the bottom, I am not!"

I first heard drummer Terri Lyne Carrington in 1979 at a Blues
Alley, Washington, D.C., teen-age jazz players' evening put together
by bassist Keter Betts. Terri was thirteen, and she performed in a
group that included Keter, tenor saxophonist Buck Hill, her father

Sonny Carrington, also on tenor, and alternating pianists John Colianni and Geri Allen. They shared the bill with the trio Pieces of a Dream.

I recall that I, along with the rest of the audience, was simply blown away by the talents of the youngsters, all of whom have become established on the international jazz scene and have albums out under their own names. I observed in my review for the *Washington Post* that Terri had the musical sophistication and chops of a seasoned player twice her age. Without question, here was a child prodigy. I was soon to learn that this was no mere subjective judgment. I had never heard of Terri Lyne Carrington before that evening, but a lot of other folks had and she already had quite a history.

An abbreviated catalogue of Terri's credits follows. She began sitting in with such jazz legends as Sonny Stitt and Betty Carter at the age of ten and at twelve was awarded a scholarship at Berklee College of Music. She recorded her first album as leader at sixteen and in 1983 shared a *USA Today* cover story, "New Jazz Stars of the '80s," with Wynton Marsalis and Ricky Ford. Two years later she was showcased at New York's Symphony Space in "The Musical Artistry of Terri Lyne Carrington," a production that was funded by a grant from the National Endowment for the Arts. She has received featured billing with her group at Jazzfest Berlin, appeared on television in duo with the late Buddy Rich, and toured Europe with Clark Terry and the Far East with Wayne Shorter. The roster of those who have enjoyed the support of her beat number Lester Bowie, Niels Lan Doky, Stan Getz, Dizzy Gillespie, Helen Humes, Mulgrew Miller, Greg Osby, Oscar Peterson, Rufus Reid, Michele Rosewoman, Pharoah Sanders, and Joe Williams, to cite only a few. In 1989 she became a member of the house band for the Arsenio Hall show. The band is known as Michael Wolff and the Posse, after its leader, the musical director of the widely viewed television talk show.

Terri can't recall when she first started hearing jazz but thinks that it was "probably in infancy because it was always playing in the house." She was attracted to the sound of the saxophone at the age of five upon hearing Illinois Jacquet and Rahsaan Roland Kirk at sessions her father took her to on afternoons in Boston, and she took up alto, playing it until the age of seven when she lost two front teeth and switched to drums.

The first set of drums Terri acquired had belonged to her grandfather Matt Carrington, who had played with Fats Waller, Chu Berry, and others, and had died the year before Terri was born. Her father "played drums a bit and showed me some basic stuff and I started taking lessons about ten all the way through seventeen," she said. Her mother, Judy Carrington, played piano and she and Terri's father were supportive of the youngster's artistic and professional aspirations early on. Her father has managed her and has directed her career. Terri credits Clark Terry with being "the first person to really give me a gig, both when I was young and when I moved to New York."

I have interviewed Terri several times over the course of the past decade and here are her recent observations on two subjects, her artistic direction as of the late 1980s and whether she believes career opportunities in jazz have substantially improved for women since she became a professional.

"My foundation is playing straight ahead, the traditional side of jazz, or whatever it is, but it's hard to do that and sound new at it because it's something that's been done for a long time now and it's been done by the greatest, and when I say the greatest I mean the innovators, and a lot of them are still alive, still doing it.

"So for me it's a real solid foundation to do damn near anything I want to do. But for me to stay up there playing traditional kinds of jazz with traditional artists, it's like living in another kind of world, because my sets of experiences are a lot different from a lot of the people I've played with. It's the stuff that I've done the longest, but I'm playing with my peers now, relating to people my own age, that kind of thing, maybe not young in age sometimes, but young in mind.

"Sure, I think it's becoming fashionable," opined Terri on the thorny issue of women's opportunities as instrumentalists in jazz. "I think some people want to jump on the bandwagon. And it's good because it's bringing more opportunities for the women that are talented. It's not good if it gets to the point—which it does in a lot of situations—just to have women just because they're women, when somebody says, 'I just want women up here because it looks good.' That's not fair for the women that have sweated blood for what they believe in."

On the rites of passage she has had to navigate in an art form dominated by males, and on an instrument nearly exclusively reserved for men, Terri quipped, "A lot of people don't expect me to be able to play that well, but I sort of have an attitude where I just want to catch them off guard, and most of the time they're shocked!"

Indeed, this Carrington admirer can attest to that very shock. At the 1988 North Sea Jazz Festival I had elbowed my way into the huge Statenhal, which seats six thousand or so in bleacher-like seats, but was unable to make any headway through the throng of a couple of thousand standees in the open space alongside the seated area. I found myself in line with the stagefront for the entirety of the closing number of David Sanborn's band. The drummer, whom I could not see, went into a rafter-shaking, several-minute-long finale solo that brought the house down. The pandemonium of the ovation that ensued made it difficult to comprehend the announcement coming through the speakers. At first, I was unable to make out the name and the phrase, but on the third or so repetition of the credits, which were being all but shouted through the microphone, I understood the garbled syllables as, "Ter-ri-Lyne-Car-ring-ton-at-the-drums!"

In 1983 the Universal Jazz Coalition/Big Apple Jazz Women came down from New York for a concert at the University of Maryland. The line-up was tenor saxophonist Willene Barton, flutist Andrea Brachfield, pianist Amina Claudine Myers, bassist Carline Ray, percussionist Nydia Liberty Mata, drummer Bernadine Warren, and vocalist Keisha St. Joan.

"I just thought it was important to get together some of the top girls to show that we are just as productive as what the guys put together," Willene explained. "We want to show that we can do it, too."

Willene early on garnered the praise of Johnny Hodges and Charlie Parker, yet throughout her more than three decades as a professional she has had to deal with the bias against female horn players. "I wanted to play music and that's all I ever wanted to do," she insisted. In addition to her membership in women's groups over the years, Willene has also been leader of many otherwise all-male combos. She recalls how audiences often looked askance when she mounted the bandstand with saxophone in hand. "It would be a matter of looking at me strange, as if to say, 'Now what is *she* going to do?' But when I started playing, you could see the changes in their faces, and there'd be smiles and handclapping and 'I told you so!' " After a recent performance a woman approached her and thanked her "for mastering the instrument I never could."

Jane Ira Bloom set her heart on trumpet when she was in the third grade, but her mother wouldn't go along with that, fearful that the horn's mouthpiece would leave a permanent indentation on the little girl's lips. "I wanted to play a shiny instrument," Jane Ira told me, "and the saxophone was the next shiniest." So she picked the alto and went on to master the soprano, which has become her principal instrument, and in the late 1980s, she was becoming recognized as one of the several most gifted players on that notoriously difficult-to-play-in-tune horn.

"When I get up on the stage and perform," Jane Ira averred, "I'm not conscious of being a *woman* saxophonist performing. I'm conscious of being a *musician* performing. That takes all my energy." Still, Jane Ira pointed out, she has been accorded the ultimate backhanded compliment, "She plays like a man," hardly an improvement on the earlier form of that style of dubious praise, "She plays good—for a girl."

Guitarist and composer Emily Remler was, in the 1980s, perhaps the best-known female jazz instrumentalist of the younger generation. Arriving on the national scene early in the decade, Emily soon found herself subbing for one or another of the Great Guitars trio—Barney Kessel, Herb Ellis, and Charlie Byrd. Other associations included vocalists Nancy Wilson and Astrud Gilberto, guitarists Martin Taylor and Larry Coryell, pianist Monty Alexander, and saxophonist Richie Cole.

Becoming something of a household name among jazz fans here and abroad by mid-decade, Emily nevertheless had to cope with the lingering prejudice against the female instrumentalist in the art form. She expressed her feelings on the double standard she said she had to contend with "every day."

Conceding that working conditions for women in jazz had improved over the course of the 1980s, Emily lamented, "But there's still a lot of things that bother me. Like people worrying about your looks when all you want to think about is the music." Emily was especially annoyed at a prominent critic who had objected (in print) to her habit of intermittently holding the guitar pick in her mouth whenever she switched to bare-finger playing. The critic confessed that he preferred to look away whenever he saw her doing this. "Good," Emily said, testily, "I wish he'd look away the whole time and picture me as John Coltrane!" Emily died while on tour in Australia in 1990. Her death was reported as the result of a heart attack. She was thirty-two.

A less sanguine view as to whether conditions have improved was revealed to me by Washington, D.C.–based drummer Roberta Washington, whose better than two decades as a working musician in a variety of contexts include playing in pick-up support groups for the likes of Aretha Franklin and Dorothy Donegan, filling the role of percussionist in pit bands, and performing in such diverse type ensembles as Latin bands, symphony orchestras, and her own combo Jazz Zen.

"It was hell and it still is," asserted Roberta. She recalled how her first teachers tried unsuccessfully to persuade her to take up clarinet instead of drums. "When I was younger all the fellows thought it was cute, but when I got to a place where I could really play and I'm a threat, then the fun stops. All kinds of subtle things take place. I'm moving my drums into a club and I'm invisible until I set them up and the other musicians are ready to play.

"Then they turn around and say, 'Man, we can't hit—where's the drummer?' And someone has to tell them, 'That is the drummer!' "

Of all the queries I have put to female jazz players on the thorny topic of opportunities for women as instrumentalists in the idiom, by far the most analytical response was that provided me by the multi-talented flute player (and flautist, one must add, for she is thoroughly versed in the "classics") Paula Hatcher.

Paula began classical flute at age eleven and started applying her skills to jazz at fifteen. One of her several records is graced by the presence of guitarist Charlie Byrd, who has also joined her in live performance. Her schedule of workshops has covered more than forty states, and in 1988 she gave a master class in jazz improvisation at the Federal University of Rio de Janeiro.

In 1989 Paula was in the studio preparing a jazz-oriented recording of works by Beethoven, Claude Bolling, and Jeffrey Baker, who plays synthesizer on the session. The project constituted for Paula a major step into the electronic realm. She played a WX7 wind controller, an instrument in the form of a computer-based synthesizer which is fingered like a woodwind and is breath-driven. "You push a button," Paula explained, "but what comes out can be any orchestral sound in the computer or what you put into the computer. Then the performer has to assume the musical character and knowledge of the instrument that has been chosen."

Paula expatiated upon the status of female instrumentalists in jazz. "I think the areas of improvement are really from the general audiences in that they will accept anyone playing as long as it's well presented and entertaining and in sync with the time. And I think as recently as ten years ago that was not the case. The last instrument I see falling is the trumpet.

"Also, in the area of improvement, I see that the male performers will perform with a woman who has it together. If you go back into early interviews with Marian McPartland, when she was starting out on her own in the 1950s, the men walked off the bandstand. Certainly those days are long gone. They do respect a woman's artistry now, and they do talk about it and so forth.

"The other area of gain is in the A&R [Artists and Repertoire] field, that if a woman can get to the producers who actually make the deal and if they're looking in terms of what the general audience is like, if a woman can get that far in the intermediate process, then she does have a better chance of being recorded than she used to have, and I think that I am beginning to see more women come out on compact discs.

"Now for the areas much in need of improvement. I think one of the basic problems that still exist—and this is an unquantifiable thing because one person's religious experience is not another person's democratic experience, and when the two become intertwined it's very hard to compete, reason, or make a move against something like this— is that many men, especially when they are young, late teens through early twenties and sometimes beyond, regard jazz playing, composition, and experimentation almost as a code or a tribal religious experience, a way of finding how they want to relate to music, to society, and they seem to feel that women have no role in this pursuit. If you look at the major religions of the world, the place for women is very proscribed.

"When young women performers who are starting out, and for whom jazz is a thing that they do rather than something from a religious perspective, come up against this, it makes it very tough for them to find enough men to work with who have a different outlook, a more, for want of a better word, commercial outlook. This explains why most groups are all-male and why the grapevine, in terms of who gets work from whom, is mostly male.

"If you take a look at musicians interacting in a nightclub, men are still the ones who can kind of sit in, talk, get absorbed with the group. A woman can sit in on a set, have a very nice discussion, but they're not cool for work. I've become used to this. I know that they don't mean it and what have you, but I've often felt a lack of being plugged into something and it does take a toll. Women, I think, are often left out in the cold and suffer from not having the interaction and fresh ideas and opportunity for the contacts.

"Now the flip side of the coin is that often you'll see with women performers kind of an iconoclast of sorts and there's a potential there, if she doesn't dry up completely, for a really unique contribution. One does have to have original thoughts and not sound like everybody else, but I believe there also comes a time when one has to interact with other people. It's wonderful to cross-pollinate with other musicians.

"I'd like to mention that there are male exceptions to the tribalism or whatever you want to call it, and there are a few who have a balanced yin and yang composition in their personalities. I found it of interest that Miles Davis uses a female percussionist. He also likes integrated bands. Another yin and yang kind of personality, I would believe, is Herb Ellis, who took Emily Remler around and said to many people, 'Look at this woman!' He did it long enough until people *looked* at her and they liked what they looked at. Another performer that I feel just kept banging at the door until finally she just knocked it in is Jane Ira Bloom, who for many years was a very well-kept secret. Finally, she's broken out on her own. So I would think that getting early experience is difficult for women. I've noticed in the last few years that the women coming to me for lessons are quite different in that they're prepared, rather than ten years ago being talented but not prepared. But now they're prepared and rather angry. Why are they angry?

"A second area of problems that women performers have is what I call the Old Territory Booker Hangover, and this is a real bottleneck. Some of the older bookers are still in the mindset from 1940 for the territorial bands that came through. They're not used to seeing women in an instrumental context, except when voice and piano are linked, if a woman sings as well as plays. Many of these bookers have residual prejudices and they use tired old lines like, 'Women aren't marketable,' and they keep asking women to provide proof of their worthiness—additional reviews, bookings, even recordings. The men aren't held to that strict standard. I don't feel bitter or anything. One has to just keep doing the best one can and persevering, because eventually one will get there.

"A third area where I think things are tough for women is that their looks are taken more into consideration than men. Let me see, they can be either anorexic or occasionally cherubic (I guess a lot of singers are that way), but not much in between. And most women and men are somewhere in between. Not everyone is an ectomorph. Some try to starve themselves. They suffer from bulimia. If a man is chiseled or sloppy looking they are regarded by the bookers as having character. If a woman is chiseled or sloppy looking she's either gay or she's let herself go.

"Another thing is wardrobe. I have a whole wardrobe, but I remember an experience when I was playing a long-term engagement in a hotel and I was asked to go in and see the head of food and beverage and the criticism was that I wore the same dress two nights in a row! It takes a woman longer to get ready for an engagement. For me to get myself in and out the door it's an hour and a half. A male performer can throw on whatever and be there in ten minutes.

"If a woman has elected to be married or have a family this can be very difficult. It means they have to juggle a personal life with husband and children, a demanding professional schedule, meals, what-have-you, because traditionally the women's responsibility has been to tend to things and, naturally, traditionally they have had a come-and-go flexibility. And on a musician's income day-care and babysitting services are almost impossible.

"Finally, there's a very sad prejudice, and to be absolutely fair I feel that I have to tell you about it. Women performers do not help other women performers. Let's say that women in positions of helping another woman, I've noticed more often then not, . . . do not return letters, . . . do not call, and I'm trying to think, why is this? And then I realized that perhaps the ultimate bottleneck of all is that the media or the attention time—that press, television, air time—is so limited for women, as if there's a percentage de facto quota how many will get air play or how many will be this or that, that the women, I think, subliminally know that they can't share the limelight with anyone else, that there's only room for one. And often you'll see a woman who has made it almost make a point of working with only male performers, when there might have been a choice.

"So I would have to say that, to be utterly honest with you. It's like tokenism, that because there's only room for so much, make sure that only a few get in. It's like they almost monitor themselves, if they have any way of doing that, and I find that *extremely* sad. But I understand it and, again, this is only a feeling that I have, but I really sense that there's something to it.

"So the thing I'm trying to do," Paula concluded her extraordinarily comprehensive overview of the obstacles confronting women as instrumentalists in the jazz world, "I'm compiling a list of all the women I've performed with who I think are good and I keep in touch with them and if there's any advice, anything I've learned to be of help,

I'll write. And in my teaching I've tried to give women performers a little extra time in talking with them in terms of things they'll have to cope with in their careers, in addition to getting things together musically and professionally. I try to let them know what they're dealing with and to help, if they want to call me or what have you."

In the early 1980s the vocalist and Baltimore native Janet Lawson, who has for some years lived in New York, was working on a musical with lyricist Diane Snow. "It'll be a musical play, but not a play in the traditional form," Janet outlined to me, "and it will be about women jazz musicians, a collage that expresses our life-style and how we picked our instruments and what kind of support we got and what it's like playing with other women, playing with men, the whole experience. At one point there's a litany that will reflect the women's names of our ancestor musicians and the music will reflect the style of their times. So it will be like the passage of the ship to this country and how the music evolved. We tried to put it into a traditional form but it didn't fit. It has to create its own form.

"Why we were motivated was that three or four years ago we were involved in a project that was about jazz and we felt that our role was missing because it was relegated to the girlfriend waiting at home and one girl who all the guys used to hang around with. There was nothing tangible about our contribution to the music. Diane and I were so upset that we decided to explore this ourselves. We had a wealth of experience and there *had* to be other women out there who did, too. We started contacting them and having rap sessions. We've just been overlooked in the history books.

"There's a great network that exists among women musicians and we are all really speaking in one voice of the feminist experience as a woman, as a jazz musician, and as a human experience in the universe. So we're really all speaking from one voice. We have to focus *in* because you have to get in touch with *your*self, but then you must come out and integrate and interweave with the rest of society, because only then can we all work together and have an impact on each other. If you stay separate I think something happens that makes the alienation still exist."

At no other time in the nearly century-long history of jazz has the art form exhibited so great a variety of approaches to the idiom as in this past decade. This is not simply because of the accumulation of styles over the course of the century, all of which are being played today, but it is also attributable to the "shrinking" of the world and the increased mobility of jazz musicians. Jazz has truly become a world music, and jazz players are among the most widely traveled artists of today.

In discussing the cumulative nature of jazz as it exists in today's world, we call upon several terms. From the broad, historical perspective, *tradition* and *continuum* are appropriate, whereas *synthesis* and *fusion* are useful when examining the various styles.

"I love it," was guitarist and bandleader Pat Metheny's unqualified endorsement of the merciless sun beating down upon us as we took our seats at an iron-framed table in the courtyard of a Holiday Inn in Virginia on an August afternoon. The mercury was close to 100° and Pat was appropriately attired in shorts and tennis shoes.

Pat Metheny and I had never before met or even talked on the telephone, but I was very familiar with his music, of which I was an admirer, and with his history, and I had reviewed him in performance on about four occasions in quite varied formats. A number of topics occurred to me as reasonable openers for an interview, but one in particular stood out as likely to elicit an interesting response. I asked Pat how he felt about the word *fusion* and about its application to what he did.

"I don't really like it," he said with conviction, "the reason being that I feel like it was a word that kind of became popular because of an ad campaign that was started in the early '70s and it's not a word that I heard musicians use at all until after that ad campaign. It really didn't come from musicians themselves. I've always felt that it was naïve to use that word as applied to improvisational music.

"When I look at the history of improvised music in this country over the last hundred years or so, the main thing that I see is that it's a music that's constantly changing and constantly being influenced by the input of the musicians and, in particular, the special things that have to do with their individual backgrounds. To me, that's what makes jazz such an exciting music, that it's a form that's willing to accept people from different philosophical, musical, cultural, and racial backgrounds and allow them to talk about something that's special to them as musicians. And to say that that's a phenomenon that just started when the electric piano was invented is ridiculous. It's something that's been there since the beginning. It has not only to do with improvisational music but from the first time, a hundred thousand years ago, that some guy in one village heard somebody banging out rhythms on a rock and then went back to his village and said, 'Hey, we shouldn't use sticks, we should use that rock!'

"Also, the word *fusion*, like the word *jazz*, like almost any of the sort of standard buzz-words that exist in improvised music, you can't find any two people to agree on what it is anyway. I hear people calling the Art Ensemble of Chicago a fusion group because they use electric bass on a couple of tunes. I mean, it doesn't really seem to have that much meaning in a sort of accepted way to anybody."

Pat's provocative observations on the issue of what is and what is not fusion triggered a follow-up question, namely, what were the elements that his music had fused, using the term in its lexical sense rather than as synonymous with "electronic jazz."

"Well, the fact that I grew up where I did has a lot to do with the way that I play now," Pat began, alluding to his Missouri roots. "First, I grew up in quite a small town called Lee's Summit that's about thirty miles southeast of Kansas City. And, really, until I was fifteen or so I had almost no contact with Kansas City at all other than the occasional trip in to see the 'A's' with my father and brother and a once-a-year trip to the Kansas City Jazz Festival that my parents always attended, being kind of jazz fans.

"I mean, it was very, like, simple. My dad had a very small business, a very nice business, my grandfather grew up and lived in the same town and his grandfather did, too. So my early days were almost like the 'Andy Griffith Show' or something.

"It was right around the time that I was nine, ten, eleven—that would be '63, '64, '65—that I started to really become a music freak, and that included everything, not just jazz, although jazz was part of it right from the beginning. I saw *A Hard Day's Night* fifteen times when I was nine. Like most kids in America, the Beatles were really something that I liked a lot. I was about eleven and my brother Mike started to bring home a lot of Miles Davis records from school and that *did* it for me. First time I heard Miles playing 'My Funny Valentine,' that whole record just destroyed me. Actually, looking back on it now, when I think that I was *eleven* years old and, like, checking out Miles every day. I mean, I'd run home from school to listen to Miles on the radio!

"Mike would have been just getting out of high school right around the time I was in junior high. He's five years older than I am. So most of my adolescence that I was learning to play, he wasn't around. So we didn't really become good friends as musicians until a few years after I had started to play with Gary Burton."

Still, musical friends or not, the older brother was early on a strong influence on the younger Metheny, for Pat took up the trumpet at eight. "My brother, who was almost like a child prodigy, started playing when he was five or six. I can remember hearing him and my dad playing trumpet duets. He showed me how to read music and what the fingering was and so forth. And my mom's father was a professional trumpet player all his life. He played under John Philip Sousa for a summer and he was also in pit bands for vaudeville groups. I can remember one of the last times that we visited him—I must have been thirteen or fourteen—having a jam session playing 'Bye Bye Blackbird' and other tunes and he could improvise great! I mean, he would call the changes. I was just beginning as a player myself and it was really amazing to me at that point to see that this guy could really play."

Pat stayed with the trumpet until braces were put on his teeth at the age of twelve. He switched to French horn in order to maintain his credit status in band throughout high school, but it was the guitar that he acquired around this time that began to claim his attention. "I remember desperately trying to find a teacher and there just weren't any," recalled Pat. And so he joined the once swelling, but now diminished, ranks of self-taught jazz artists. And he listened.

"Even though Kansas City was relatively close, Lee's Summit was really a different place. I mean, country music was the thing. You couldn't walk out the door without hearing Patsy Cline and Porter Wagoner and Dolly Parton. That whole thing was very prominent as the local music. And the fact is, my parents *hated* that music, couldn't stand it. It just so happened that there were some neighbors that we had that really loved it. I used to go over to their house and they were always there and there were always guitars there and they were always playing them. I mainly watched, I didn't really play that much at that point, and it really fascinated me. I was more or less used to hearing one or two guitars, but there were seven or eight guys playing individual parts that were different from each other, like in a bluegrass kind of format. It influenced me more as a listener than as a player. As far back as I can remember, that music was just in the air. It fit the territory so well.

"My whole relationship to the blues was strictly from the sort of Jimmy Smith, Kenny Burrell point of view, starting from when I was about ten when I started to get their records, and even to this day Kenny Burrell is by far my favorite blues player. Wes Montgomery, Jim Hall, and Kenny Burrell, those were my idols. Especially the first year that I played seriously, Wes was the cat. I didn't use a pick at all. The first year I played only with my thumb. I could do a pretty credible Wes imitation."

Pat Metheny had already done a lot of listening when, at the age of fifteen, he "got the chance to work five or six nights a week in Kansas City with *great* players, I mean, the best players in town." Crediting these musicians as his first on-the-job teachers, he describes their approach as mainly straight-ahead. "We played mainly standards and I remember we did a lot of Freddie Hubbard and Stanley Turrentine tunes. The other aspect of it was the Kansas City thing—we did shuffles all night long. I was so lucky to get to play with these guys."

The first big break for Pat on the national level was with Gary Burton's quartet. Already an ardent admirer of the vibraphonist by the late '60s when he was playing those Kansas City gigs, Pat was buying every Burton album as it came out. "I would transcribe all the tunes and learn to play them and force all my friends to play them with me. That band was so exciting to me because it was really the only band in jazz at that time that wasn't led by a guitar player where the guitar had a really major role in the music. The way that Gary set up his group at that time where all the instruments were really equal was something like a chamber group, but they were still playing on changes, [playing] things that were really descendants of the Bill Evans Trio with Scotty LaFaro and Paul Motian.

"When I finally got to meet Gary and to play with him I was scared to death. I literally remember seeing my knees shaking on that occasion. This was at the Wichita Jazz Festival in, I think, 1982. Gary could easily have said, 'Who is this jerk kid who's like drooling all over me?'

But he saw whatever he saw in my playing and took me under his wing.

"After talking about how important he is to me as a musician, maybe more important is that he took me from being from this little town in Missouri to the East Coast, showed me what the music business was about, described to me in detail everything about his own career, the way his operation worked. I think he knew that I was going to have my own band at some point and that I needed some help just to kind of figure out what was in store for me. I feel like my band's music is very much a descendant of what his band was doing, particularly in the late '60s and early '70s."

The summer of our interview marked the tenth anniversary of Pat Metheny's band, and his association with keyboard player Lyle Mays, who had been with the band since its inception, pre-dated that by three or four years. Besides his three-year membership in Gary Burton's combo, Pat has had strong associations with saxophonists Sonny Rollins, Dewey Redman, and Ornette Coleman, bassists Charlie Haden and Jaco Pastorius, pianist Paul Bley, drummers Billy Higgins and Jack DeJohnette, and others.

That Pat Metheny's music is fusion—but in the dictionary sense of the term, "the merging of different elements into a union"—is clearly the result of the many influences that have shaped his artistic ethos. In addition to the idioms and musicians referred to up to this point, Pat added the following as among those he has listened to and admired since becoming a "music freak" some quarter-century before we talked. There were Louis Armstrong, Lonnie Johnson, Django Reinhardt, Charlie Christian, Lester Young, Art Blakey, Roy Haynes, Philly Joe Jones, Wilbur Ware, Clifford Brown, Art Farmer, Grant Green, John Coltrane, Paul Chambers, Ron Carter, Pete LaRoca, Freddie Hubbard, Joe Henderson, McCoy Tyner, Keith Jarrett, Herbie Hancock, Chick Corea, Tony Williams, Jerry Hahn, Larry Coryell, James Blood Ulmer, and the composers Ravel, Stravinsky, Steve Reich, and John Adams.

"To me the biggest problem in jazz right now is that there aren't a lot of new musicians showing up that are really quality players," Pat continued. "It's very disturbing to me, to tell you the truth. I was so excited when Wynton Marsalis came on the scene. It was the first time I heard somebody behind me chronologically who could *really* play, could really play on harmony, and understood the role of his instrument in jazz. I mean, most of the so-called 'fusion' guys, almost all of the avant-garde guys, almost all of the world music guys don't understand harmony, and this bugs me to death. It's because they haven't worked on it. It's not easy, it's not easy at all.

"The general lowering of standards in jazz bothers me. I pick up the jazz charts and I don't even recognize most of the names on it. And I'm saying this, wholly aware that I'm one of the cats that's on the charts playing an electric instrument in a controversial setting. The difference, for me, is that I can honestly say, without feeling like I'm being egotistical: I know what I'm doing. I'm aware of where our thing is and what fits into the scene.

"I know that what I play has a relationship to what's happened before. And that just doesn't exist with most of those other names that I see on the charts, and that really bothers me. I mean, when I see Wynton making these very, like, strong statements about standards and everything, I'm behind him one hundred percent. I think that people should be playing a lot better than they're playing and young players should be taking the music a lot more seriously than they do.

"The player that I feel the closest to who is furthest away from me is Django Reinhardt," Pat pointed out. "Prior to that, I have to admit, it's distant from me. Of earlier guitarists, I know Lonnie Johnson's playing the most. This is an area that I feel sort of bad that I don't know more about, and when I hear it I think, man, it's amazing that those guys were playing like that then, and this is particularly true when I hear Louis Armstrong, it sounds so modern and fresh.

"The thing is, when you're fifteen to twenty there's something about being able to identify with the musicians," Pat conceded. "I see it happening now with young guys who come up and say, 'Oh, you're my favorite guitar player!' And I say, 'Well, have you ever heard Wes Montgomery?' And they say, 'Who?' And I want to kill them!"

The jazz idiom has had its multi-instrumentalists, for example, Benny Carter and Ira Sullivan, who play reeds and trumpet, Anthony Braxton and Vinny Golia with their veritable arsenals of saxophones, and the Brazilian Hermeto Pascoal, who is self-taught on instruments representing the reed, woodwind, brass, string, keyboard, and percussion families, and he sings, to boot. New Orleans, in the early years of the century, had Manuel Manetta, an associate of Buddy Bolden. Manetta performed on and taught all instruments and was known for his ability to play trumpet and trombone simultaneously and in harmony. Sidney Bechet established a recording first when he overdubbed soprano and tenor saxophones, clarinet, piano, bass, and drums for his 1941 "Shiek of Araby" 78 rpm. Rahsaan Roland Kirk often blew three saxophones at the same time. Native Australian James Morrison plays brass, reeds, and piano.

In the mid-1980s a self-produced album, *Multiple Instruments*, featured Scott Robinson on more than thirty reed, woodwind, brass, and percussion instruments. In addition to the expected—trumpet, trombone, saxophones—Scott performed on such esoterica as Highland bagpipes, ophicleide, and "solaristic sound sculptures," metal sculptures played with a cello bow. A collector of instruments, both antique and modern, Scott has acquired over a hundred, most of which he has mastered, including a bombarde, rotary valve posthorn, normaphon, mellophone, C-melody and bass saxophones, and valve trombone, to name a few that one does not often see on the bandstand.

"For me it's really a fascination with sound," explained Scott, adding that his listening habits had long encompassed a spectrum from Louis Armstrong to Sun Ra. "As a composer and as an instrumentalist, when I hear a sound that I like from a particular instrument, right away I have ideas for where I can use that sound. It's amazing how many sounds you can achieve with just acoustical sound production sources. You know, I haven't gotten into synthesizers. To me it's too much fun fooling around with acoustical sounds. I had a friend at Berklee School of Music, Emil Viklický, a wonderful Czechoslovakian pianist, and he used to say, 'You're the living synthesizer.' That was a great compliment."

Thus we have, in the first two artists of those chosen as representative of the current scene, quite different approaches to the element of sound and the sources thereof. One chooses electronic sources, the other acoustic. Yet both are, in the dictionary sense of the word, committed to the concept of *fusion*.

"So a man plays differently from the way you do," cornetist Warren Vaché remonstrated in a discussion with me. "Why all the defensiveness?" Warren, believed by some to be the heir apparent of Louis Armstrong, Bix Beiderbecke, Rex Stewart, and others who pioneered the classic jazz trumpet styles of the 1920s and '30s, was reacting to the criticism, from some of his elders, of his admiration for such modern jazz artists as trumpeters Fats Navarro and Clifford Brown. Warren has worked with Benny Goodman, played the musical part of Beiderbecke in repertory programs, and appeared in the lead role of the 1984 film *The Gig*, which relates a Dixieland band's two-week stand in the Catskills.

Expressing a love for the music of Charlie Parker as well as for that of Armstrong, Warren explained that "Each strikes some sort of emotional chord in my ear. It's music—what's the difference? So the idea for me would be to come out with some sort of synthesis. I've got to work at it. That's what it's all about."

The three musicians from whom we just heard were born in, respectively, 1954, 1959, and 1951. Another artist, born in 1936, who well illustrates the concepts of synthesis, tradition, continuum, and fusion is bassist Buell Neidlinger. In fact, Buell has to his credit perhaps a more varied and disparate roster of musical associations than any other player in the history of jazz. They include, to name only a representative few, Willie The Lion Smith, Wild Bill Davison, Billie Holiday, Joe Sullivan, Pee Wee Russell, Sonny Greer, Vic Dickenson, Coleman Hawkins, Dick Wellstood, Ben Webster, Tony Scott, Zoot Sims, Clark Terry, Tony Bennett, Freddie Redd, Steve Lacy, Roswell Rudd, Gil Evans, Dennis Charles, Don Cherry, Archie Shepp, Roy Orbison, Ry Cooder, Cecil Taylor, Peter Ivers, Peter Erskine, Barbra

Streisand, Jean-Luc Ponty, and Frank Zappa. He has also been a member of the Boston, Houston, and other symphony orchestras, and of the Los Angeles Chamber Orchestra, has concertized with the Budapest and Amadeus quartets, and "had the very first band to which the words 'jazz-rock' were ever applied." Since taking up residence in California in the early 1970s Buell has "been very busy in the Hollywood studios," he told me. "And of course I've always had a band and played the music," he added. His bands have included Buellgrass, Thelonious, and String Jazz.

"It's always been my philosophy," Buell continued, "that to be a complete musician, one has to play all styles and be able to play in a hot band or a symphony orchestra."

Buell's combo Thelonious is exclusively devoted to the Monk repertoire. The fare of String Jazz, which uses tenor saxophone (Marty Krystall), violin (Brenton Banks), mandolin (John Kurnick), drums (Billy Orborne), and the leader's bass, had recently completed work on an album, *Locomotive,* for the Italian Soul Note label. For the session seven Monk tunes and four by Ellington were chosen.

"You see, what I feel is, Ellington and Monk are basically the same person in terms of their *oeuvres,*" Buell insisted. "They both referred to what I call the New York style of piano playing and that's where they come out of. When one says that Duke Ellington is a swing musician and Thelonious Monk is a bebopper, one shows a great inability to judge musical styles.

"What I'm trying to do is preserve the work of these guys, in a way, and expand upon it. We accept the forms that the masters gave us and we try to use them in terms of today. So we're interested both in preserving the forms and at the same time putting them into the context of today's playing."

What other player in the jazz idiom can claim as his own such disparate musical experiences as having been "on every recording that Cecil Taylor made for the first seven years of his recording career" and such on-the-job training as that illustrated by the following anecdote?

"Sonny Greer said one of the most amazing things to me once at the Metropole when I first came to New York," Buell recalled from the early 1950s when he was still in his teens. "I worked with a band that included Sonny and there was a dressing room at the Metropole, but Sonny had found a large refrigerator box and had moved it upstairs on this little loft balcony and that was his office, as he called it. After the first set he said, 'Come up to my office.' I said, 'Okay,' and I went upstairs and he said, 'Do you have a mother?' I said, 'Oh, yes, Mr. Greer, I do.' He said, 'I bet she's a very nice lady,' and I said, 'Yes, she is,' and Sonny said, 'Yes, I had a mother, too. She was a very nice lady.' He said, 'Tell me, did your mother ever tell you that it was rude to rush?' Relating this to me, Buell laughed at this point. "It was one of my lessons in music from a master like that. What a way to put it, huh?"

Don Cherry, who was born in Oklahoma City and grew up in the Watts district of Los Angeles, has expressed admiration for the early cornet work of Louis Armstrong and Bix Beiderbecke, and in the 1960s he played an important role in the seminal free-jazz groups of Ornette Coleman. In the 1970s he became the virtual leader of a new pan-cultural school of jazz which he and others call "world music." His chief instrument is the "pocket" trumpet, but he has immersed himself in the musics of India, Africa, China, Tibet, and other cultures and has mastered bamboo flute, the bowed calabash, the "hunter's guitar" of Mali, and the doussn' gouni, a long-necked string instrument with pumpkin-sized gourd for a sounding box.

In 1983 Don Cherry came to Washington, D.C.'s Blues Alley for a performance with the trio Codona, the other members of which were the Brazilian percussionist Nana Vasconcelos and New Yorker Collin Walcott, who had studied sitar with Ravi Shankar. (He died in 1984 at age thirty-nine as the result of a bus accident.)

"The naturalness of music is what I want to be involved in," Don told me. "It's like the instruments are playing us, and I can really see now that it's time for that because of the acoustic-ness and all the instruments are working with overtones and the idea is the swing in it and we're really creating a nice boogie, you know."

Don pointed out that the group was not playing the traditional music of the several cultures they had delved into but writing originals and using tunes of Ornette Coleman and others, "and playing them with these instruments." Vasconcelos's myriad approaches to rhythm utilized shakers, drums, berimbau (a bow tapped with a stick), and body percussion. "This wave, that wave," Don enthused. "I feel now it's an *all-wave* music that's happening!"

The jazz idiom boasts few players who have spread themselves as widely across the musical horizon as has the Texas-born guitarist Larry Coryell. As a teenager in the late 1950s in Seattle, where his family had relocated when he was seven, Larry played in a rock and roll group.

Having settled in New York in the mid-'60s, Larry was a member of the combos of drummer Chico Hamilton and Gary Burton. In the '70s his associations included Charles Mingus, Sonny Rollins, Herbie Mann, Stephane Grappelli, Michal Urbaniak, and Miroslav Vitous. He has performed in duo with guitarists John Scofield, Vic Juris, Steve Khan, Philip Catherine, and Emily Remler. In 1973 Larry formed Eleventh House, an early jazz-rock group, with trumpeter Randy Brecker, drummer Alphonse Mouzon, and others.

In the 1980s Larry found himself working with Claude Bolling's combo and in the trio Guitarjam with Brazilian native and Stan Kenton band alumnus Laurindo Almeida and Sharon Isbin, a young classical guitar virtuoso from St. Paul, Minnesota. In both of the just named formats Larry was playing classical, Brazilian, flamenco, and jazz. He was also performing and recording with his own combos of a straight-ahead nature, and toward the end of the decade he was planning to put together a new formation of Eleventh House for concert appearances. While it is difficult to categorize Larry Coryell within any one, or even several, styles, two aspects of his playing are undeniable. He is a supremely gifted improviser and a master of his instrument.

"I've always been kind of involved in one kind of fusion or another and this is yet another kind of fusion" said Larry, explaining for me his collaboration with Almeida and Isbin. "It's a healthy departure from the straight-ahead jazz format I've been doing as of late. It's mainly a classical gig where for me the emphasis is on discipline rather than improvisation, you know, playing my parts properly and integrating with the group. I have a solo, usually, take a classical theme and improvise on it, and sometimes Laurindo and I will do some Brazilian stuff, like straight-ahead bossa nova, and we'll both improvise there. The only one who doesn't really improvise is Sharon, of course.

"I just like to do a lot of different things, I like a lot of different kinds of music and always have. I wouldn't call it diffusion so much as versatility. Every type of music that I do requires a total amount of concentration in order to do it right, starting with the thing I do with Laurindo and Sharon and, for another example, the things that I do with Claude Bolling. That requires a tremendous amount of mental work in order to make it happen right. I like to do a lot of different things and I like to try to do it as well as I possibly can. A long time ago someone who I have a lot of respect for, Charles Lloyd, told me never to limit myself, even though I would always come from a jazz foundation, to try to do some different stuff, keep doing different things, not only for my own benefit but the benefit of the listener."

It intrigued me that Larry had played both high-decibel electric guitar in rock and rock-jazz formats and nylon-string classical guitar as a soloist, and so I asked him to talk about the differences between the two types of instrument, pointing out that he had in the past remarked upon the more sensual feel and sound, for him, of the acoustic guitar.

"Well, it just happened to work out that way, Royal," he began. "I know a lot of straight-ahead electric jazz guitar players who get the same sensuality from their traditional jazz guitar-cum-amplifier, but just the way I work, it seems more emotion gets transferred through the acoustic.

"The electric guitar in the context of a more or less loud thing was really a horse of another color, because you get so much more power from the note, each separate note has got all this electricity behind it, it's got a lot of wattage behind it, and it really becomes another instrument. You have to change your touch, you have to be more aware of how your settings are, et cetera.

"When I switch to the electric guitar these days I play it with a *much* lower volume and a much more traditional sound, like the sound that attracted me to jazz guitar in the late '50s, you know, like through a relatively small amplifier and no overloading or distortion. In the late '60s I sought a certain creative innovative outlet with distortion, and other players also picked up on that and took it a lot further than I did. But now I'm just trying to get more of a pure sound and I use very few effects when I play an electric guitar. For example, on the record with Sharon and Laurindo I play a Gibson 400 through an amplifier on two songs at a very low volume."

Larry had alluded to his attraction to jazz guitar in the late 1950s, but no names had been cited. I asked him who some of his idols were, as far as guitar players.

"I'd have to show a debt to Tal Farlow, a big one, and Barney Kessel, and I love Johnny Smith, and I really, really loved Grant Green. With Emily"—he had recently been touring abroad with Emily Remler—"I've been exposed to some of the things she learned from people like Joe Pass and Pat Martino, and I'm re-examining those people through her eyes. I already had a deep appreciation of them, but it's a much deeper appreciation because she really has their styles down." I reminded him that he had in the past included Charlie Christian and Django Reinhardt, to which he responded, "Oh, I forgot to mention those guys!" When I asked whether he went back into the music's history to check out such pioneers on the instrument as Bud Scott and Lonnie Johnson, Larry confessed, "Yeah, but not nearly enough as I should because I'm so busy out on the road."

My final question to Larry was along the lines of how did he view the present scene, did he think that the high visibility of the guitar in the music was attracting young people to jazz, and what would be his own direction in the future?

"As far as I can see, a definite affirmative on that," he enthused. "I've never seen such a proliferation of young, pure jazz guitarists and guitar fans coming up as I have lately. A lot of it has to do with people like Emily. Pat Metheny had a lot to do with it, and people like John Abercrombie, John Scofield. It's really great and you can get a pure jazz audience, albeit a small one but still a loyal one and a very enthusiastic one, any place on this globe today. I think the guitar has a lot to do with it.

"A lot of us have gone back and re-examined the importance of Wes Montgomery in our formation and our roots," Larry said in conclusion to our discussion. "I'm harkening back to a lot of that and it sounds good now, whereas when I was concentrating on the fusion thing in the middle '70s I just didn't have time to hone the basic improvisational skills that came from Wes Montgomery, that school. I mean, I was involved in ensemble playing, fiddling around with electronics, learning how to use electronics, trying to get the best sound from them. When you take away all those electric accoutrements you're left with going back to honing the improvisational skills, which never really change.

"I'm concentrating on just doing that because those really are my roots, that's what got me into it in the first place. With my rather mercurial thinking patterns I've gone through a lot of styles and not wanted to limit myself, but nevertheless my roots really started there and now I'm going back and reconnecting with those roots and I'm finding that they're very deep roots, there's a lot in there, you can do a lot of things.

"It has nothing to do with how many records you sell, how many people come to hear a concert, it's a concern with honing the basic improvisational skills and working over the materials that people like Wes Montgomery were making such great inroads with in the '50s and early '60s."

We surveyed the big band scene from its formative years to the present in an earlier chapter. Two big bands that were not discussed and which will serve as representative of the big-band scene in the 1980s are those of Charli Persip and Sun Ra.

"I'm the kind of drummer and musician that I try to stay in touch with what's going on all of today," said Charli Persip, leader of Superband. "I really try to deal with what's happening. Superband plays music as it is our conception of big-band jazz of today, going *into* the future."

There is no other big band drummer-leader playing today who can back up that kind of back-to-the-future musical philosophy with the combination of years on the scene and open-mindedness that makes Charli Persip unique. For it is not simply that Charli's musical memory reaches back half a century to Apollo Theatre matinées that he attended with his father and which exposed him to, among other legends, drummer Cozy Cole ("I was mesmerized by the whole stage thing," he recalled in wonderment. "Oh, my goodness, that made a tremendous impact on me"). It is, additionally, that Charli is a musical visionary who surrounds himself with young players whose raison d'être is artistic daring. It is a case of wedding respect for and knowledge of the jazz tradition to a risk-taking spirit that makes Charli as individualistic an artist as we have at the helm of a jazz orchestra today.

Charli Persip's drums have provided pulse for, among others, Gene Ammons, Tadd Dameron, Eric Dolphy, Harry Edison, Don Ellis, Frank Foster, Red Garland, Quincy Jones, Rahsaan Roland Kirk, Hal McKusick, Hank Mobley, Lee Morgan, Johnny Richards, George Russell, Archie Shepp, Zoot Sims, Dinah Washington, and Phil Woods. He spent five years in the drum chair of Dizzy Gillespie's big band in the

1950s, was with the Harry James band briefly, and recorded with Gil Evans on *Out of the Cool*.

Charli has had as sidemen in his own small combo such notables as Freddie Hubbard and Ron Carter, performed in trio format with Eddie Gomez and Jack DeJohnette, been musical director for Billy Eckstine for seven years, and served as drum instructor in the 1970s for Jazzmobile in New York. He has led drum clinics here and abroad, appeared on prime-time television talk shows, and published a book, *How Not To Play Drums*. Indeed, Charli Persip has been there and back.

"Our band is trying to plug into the *true* jazz concept where everyone in the band knows the tune," Charli explained. "Not only do they know the arrangement, they know the *tune*. So consequently our band will have the feeling of a small band, because in a small band nobody has to be told where to come in and so forth and so on. Instead of depending on me to tell them when to come in, they use me as an inspiration. In other words, my time feeling is inspiring the band to bring out their own time feeling. Consequently we have a much looser feeling in the band and I'm able to play loose, as I would in a small band. Drummers shouldn't sound any different in a big band or a small band, except maybe for volume."

The great drummers of jazz have necessarily been attentive listeners, whether in combo or big band format, and the very greatest of them have this talent to an uncanny degree. I asked Charli how he acquired his highly developed ability to keep the beat going and at the same time not only listen to but respond to—frequently to the mere nuances of—what a dozen players are doing around him.

"I trained myself to do that," Charli answered without a moment's hesitation. "Not only when I started playing but when I became a little more mature and kind of knew what I was doing. I started aiming myself in the direction of really trying to listen and I became really aware of how important listening was for drummers, because we play an instrument of accompaniment and we must listen so that we can accompany properly and at the same time still be creating. That's really basically what I've been trying to do, trying to perfect what I call the art of creative accompaniment."

Of all the colorful characters of jazz history, surely keyboard player, bandleader, composer, and self-styled mystic Sun Ra would have few rivals in a competition for first place as the most outrageous. Born in 1914 in Birmingham, Alabama, Herman Blount's early career included membership in the orchestras of Fess Whatley (mid-1930s) and Fletcher Henderson (1946–47), playing piano gigs in Chicago in the late 1930s, associations with Stuff Smith and Coleman Hawkins in the 1940s, and formation of his own trio in the early '50s with saxophonist John Gilmore and bassist-saxophonist Pat Patrick, both of whom would become members of his big band and remain so into the '80s.

Herman "Sonny" Blount adopted the name Sun Ra and dubbed his band the Arkestra before moving from Chicago to New York around 1960. The 1970s saw him settling in Philadelphia, where he has kept his Arkestra intact in a communal living arrangement. He has recorded many albums on his own Saturn label, toured the world, and been documented on film. Ra's sidemen have included trombonists Julian Priester and Craig Harris, saxophonist Marshall Allen, bassist Ronnie Boykins, and drummers Lex Humphries and Clifford Jarvis.

A brilliant showman, Sun Ra incorporates into his performances such elements as dance, multi-percussion, chants by full band, audience mingling, slide shows, extravagant costuming of the orchestra on ancient Egyptian, outer space, and futuristic themes, and pseudo-philosophical monologues. Some critics, offended by Sun Ra's bizarre stage presence and his seemingly pompous pronouncements from on high as to his extraterrestrial origins and divine status, close their ears to his music, which is, after all, the core of his art and of no little significance in the evolution of the music from the 1930s into the '80s.

While contemporary jazz has always been a decade or two ahead of popular taste, even its most forward-looking creators have characteristically made their statements in musical terms alone. Not content with that, Sun Ra long ago assumed the tripartite role of messiah, mythographer, and prophet of, as he would have it, Afro-Universal artistic expression. Given the opportunity, Ra will expatiate upon his fantastical adventures with aliens who, he insists, have beamed him up to their celestial habitats for seminars on the future of Planet Earth.

I was granted an audience with the Ra a few years ago at evening's end (of a single, four-hour set, that is) during a several-night engagement of the Arkestra at the Show Boat in Silver Spring, Maryland. Our interview began at about 1 a.m., and we sat and talked into a tape recorder into the early hours of the morning on a variety of topics. The discussion gave me a great deal of insight into this artist's work.

"Music represents the potential future," Sun Ra began. "It's about the future that's not supposed to be but that is better than is supposed to be. Things today are not really suitable for our better world, so therefore we need something else and music is a sort of advance herald of something other than what is. Therefore my music is presented through myth and through what I call cosmic drama. People can understand better on such a vast plain as what I'm speaking if it's presented through the dance, 'cause dance is music, too, and presented through vocal, because that's music, too. Silence is also music.

"So all of it's put together, each in its proper place, as a language expressing things concerning the potential future, which I call the alter-destiny, realizing that Earth is getting over into what you might call different isolated armed camps, and if people don't get something in common quick, well, they're not going to have anything at all."

The term *freedom* is routinely used to describe the music of Sun Ra as well as that of others in the avant-garde of jazz, but Ra has strong views on the subject. "Musicians, if they're going to play in an orchestra, they *have* no freedom, they have to be disciplined, above all other men on the planet. Other men, they can get by, but musicians got to hit the right note, they got to be in there, got to be in tune. A musician is really the model for men who want to be in accord, in the way *accord* got the word *chord* in it, you see.

"Musicians really represent the harmony department of the universe. The fact that a lot of musicians compromise and they're doing other things mean that they don't understand their mission up on Planet Earth. It is to really *be* together, then when other people see musicians together, they will get together, 'cause men like to imitate one another. A band together can inspire armies to be together, can inspire children's schools to work together, can inspire social workers to work together, it can inspire everyone to be in harmony, be in accord."

Ra's love of metaphor is evident in his observations on his audience. "People are just like receivers, they're like speakers, too, like amplifiers. They're also like instruments because they got a heart that beats and that's a drum. They've got eardrums, too, and they got some strings in there, so they actually got harps on each side of their head. If you play certain harmonies, these strings will vibrate in people's ears and touch different nerves in the body. When the proper things are played for each person, these strings will automatically tune themselves properly and then the person will be in tune. There will be no discord, they will be tuned up perfectly, just like each automobile have to be tuned according to what kind of automobile it is. My music actually does have a vibration somewhere within it that can reach every person in the audience through feeling.

"In Africa we were rehearsing and a lizard came up and stayed by the speaker for about thirty minutes. In France a dog lay on stage for an hour and a half while we were playing, curled up in front of the organ. They say that music soothes the savage beast. It can reach animals, and it can reach the most dangerous animal of all—man. That's what it takes to reach man and make him feel something about his brother. There's a possibility that this music can do that, and if they keep on listening, they will understand and the music will harmonize them to a point that a light will glow in their mind and they can see themselves. If they're pure in heart, they won't see anything but the purity in their hearts and those kind of people are the ones that will feel the music first, and through them it will be continued, 'cause happiness is just as contagious as anything else."

It was the music he heard as a youngster that formed his standards of judgment, Ra pointed out to me. "I grew up as a child not really understanding what I was hearing," he explained. "My people liked the theater and they always took me to see what I now realize was stars, like Bessie Smith, Mamie Smith, Clara Smith, Butterbeans and Susie. Every week they would take me to see some people like that, people who are really giants. I got a chance to be in on the beginning of something.

"Therefore my criteria comes from my early childhood of hearing nothing but masters. I didn't know anything else and I can tell the difference. Also, they happened to have a lot of Fletcher Henderson's records and I always played them. I mean, I didn't even know the name of the band, 'cause I was hearing it from the time I was two years old. Later, in high school, I began to distinguish between bands and I began to listen to Duke Ellington, Fats Waller was in there, too, then Art Tatum, then Jay McShann. I was very interested in each man and looking for something different."

Indeed, Sun Ra's Arkestra blends a great many different elements. There are, for example, Ellingtonian voicings, Henderson-like structures, and solos that recall Lester Young, Cootie Williams, Charlie Parker, and Charlie Christian. The leader himself will play backroom blues here, Count Basie economies there, with the band riffing behind him à la Jazz at the Philharmonic. Then there are updates of the foregoing, for notwithstanding Ra's strictures on the importance of discipline, his players can blow free with the best of them.

"I had to teach my musicians an entirely different concept," said Ra to me in the by now nearly dark club. "Just like, if you want to speak in a different language, you've got to teach the people the grammar of the language. So I've been teaching them the grammar of the music."

Each generation of jazz artists produces players who emerge early on as the most likely to step out in front as major voices of the idiom, archetypes of the dominant style of the period, perhaps even innovators. Few observers of the 1980s jazz scene would question the candidacies of saxophonist Chico Freeman and trombonist Craig Harris for membership in that select group out of the pool of musicians who are several years into their second decade of professional activity as we enter the 1990s.

With fifteen or so albums under his name and having contributed to another twenty or more, multi-reed and woodwind player Chico Freeman has compiled quite an impressive roster of associations. To name a mere representative few of them, Chico has worked with Muhal Richard Abrams, Mickey Bass, Jack DeJohnette, John Hicks, Jay Hoggard, Bobby Hutcherson, Elvin Jones, Jeanne Lee, Steve McCall, Wynton Marsalis, Don Moye, Don Pullen, Sun Ra, Sam Rivers, and his father the saxophonist Von Freeman.

With his longtime musical companion the bassist Cecil McBee often present on the date, Chico has recorded salutes to the standards, flat-out blowing sessions, syntheses of jazz and African idioms, contemporary on-the-edge expression, and brilliant summings-up of his multi-form approach to the art form of jazz. In the mid-'80s Chico collaborated with the British virtuoso John Purcell in a release titled *The Pied Piper,* on which he doubled on seven horns and Purcell on five. A prolific composer, Chico writes tunes for virtually every album he appears on.

Chico's exposure to music came early and it was constant, what with growing up as the son of a saxophonic stalwart of the Chicago scene. Then there were his uncles, drummer Bruz Freeman and guitarist George Freeman, who often practiced, along with Von, in the family home, where 78 rpm records were nearly always spinning on the turntable. Picking out melodies on the piano, Chico had written his first tune by the age of seven. Coming across a trumpet in his father's gear one day, Chico took up the horn and played it through college, adding saxophone along the way.

Some of Chico's earliest performing experiences were in Chicago rhythm and blues combos, and he soon hooked up with the Association for the Advancement of Creative Musicians (AACM), coming under the wing of Muhal Richard Abrams. Chico ranks Abrams's influence on him as equal to that of Ellington. Then came undergraduate studies at Northwestern University and, in 1974, enrollment in the masters program in composition at Governors State University. Chico received the award for Outstanding Instrumentalist/Best Saxophonist at the 1976 Collegiate Jazz Festival, held at Notre Dame University (and judged that year by Lester Bowie, Bob James, Malachi Favors, Joe Farrell, Don Moye, Dave Remington, and Dan Morgenstern). That same year Chico toured Brazil as a member of GSU's big band and soon after returning to this country he relocated to New York.

A decade after moving to New York he did not find himself spending much time in his adoptive city. When he is not on the international road with the now five-year-old group The Leaders, Chico finds himself touring here and abroad with his father or with his own combos. Then there are the educational ventures he is involved in, "traveling around the country, actually around the world, conducting clinics and workshops and lectures from grammar school to high school and colleges and universities." He spoke of plans for several new albums, including one to be dedicated to Duke Ellington and containing compositions "within a Duke feeling," and of a new group he was putting together, hinting that it would be a departure from what he had been doing of late.

"One of the great things about being a musician," Chico enthused, "is that I have the opportunity to travel all over. I've been to Europe, eastern Europe, Australia, Japan, the Caribbean, all over the U.S." At a festival in England Chico found himself performing before an audience of 20,000. "I have found a great reception overseas," he said, "more so than in the U.S. You go to a place where people are of one language and one culture, but yet they're very open to what you're bringing them.

"I find that the most rewarding kinds of experiences that I've had," observed Chico, our discussion having turned to his membership in the star-studded sextet The Leaders along with Lester Bowie, Arthur Blythe, Kirk Lightsey, Cecil McBee, and Don Moye, "is when all the people involved allow the music to dictate what's to happen. It feels as if you really had almost nothing to do with it because you take it out of your mind and put in into your heart and spirit. When you do that, everything you know is there anyway and you choose the right things almost automatically. It's like going to a computer and picking it out. It's there and it comes out and it meets the call when the call is there.

"I think that this music," Chico summed up his thoughts, "the love that's put into it, the freedom that it represents, in the sense that the whole idea of playing it is about freedom and throwing it into one pot so that all the people involved can work together, that's the idea of this music. This music, in essence, is the epitome of what people are speaking of that they want in their lives—freedom, democracy, the ability to say what they want to say when they want to say it, but at the same time mindful that we have to live with each other. I think that's probably the biggest contribution that this music has given to the world."

Few in the wider jazz community had heard of trombonist Craig Harris until he abruptly surfaced in the jazz polls in the mid-1980s. He had worked with Sun Ra and Abdullah Ibrahim in the mid-1970s and late '70s, respectively, and by the early '80s had formed associations with David Murray and Muhal Richard Abrams. Subsequently he has played in groups led by Lester Bowie, Henry Threadgill, and Olu Daru, and he has been a member of Charlie Haden's Liberation Music Orchestra and Jaki Byard's Apollo Stompers, as well as big bands led by Sam Rivers, Cecil Taylor, and Abrams. In the 1980s Craig was performing and recording with his own groups, including Tailgater's Tales, a quintet with instrumentation along the lines of an early piano-less New Orleans band.

"*Every* decade has been the decade of the trombone," was Craig's response to my suggestion to him that the 1980s were going to be the instrument's decade, which I noted had been a long time coming. "There've always been good trombone players." Still, even as loyal a trombone partisan as Craig had to concede that the horn had been somewhat out of fashion during the 1960s and '70s, the saxophone having held sway during that period. Craig offered several explanations for the trombone's decline in popularity during those two decades.

"What J.J. Johnson laid down in the '40s and '50s was so awesome, it took people a long time to even try to do it any other way. His standard was so high that anybody that did not play like that probably did not work. If they couldn't find a J.J. Johnson, people would say they didn't need a trombone." The result, Craig averred, was that the quartet fronted by a saxophone, "maybe a trumpet," became the typical jazz combo format for years. He also pointed out that composers were not writing for trombone during that period.

However, all that was changing, insisted Craig, and "a great group of young trombone players is coming up and they can do anything on the instrument, no restrictions." In the forefront of this new generation of trombonists are, we might add, such players as Ray Anderson, Dan Barrett, Robin Eubanks, Frank Lacy, George Lewis, Delfeayo Marsalis, James Morrison, Steve Turre, Gary Valente, and Harris himself. They are virtuosi to an individual, and a large slice of the jazz trombone tradition can be detected in the playing of each.

Craig attributed his own catholicity of musical sources to his four years at the State University of New York at Old Westbury, where "they had an Afro-American music program, and it just opened my whole horizon to this culture."

As far as his early life is concerned, Craig Harris's has to be one of the more uplifting success stories in the annals of contemporary jazz. "I grew up in project houses," said Craig, recounting his youth in Hempstead, Long Island, "and all my life I had been hearing music, a lot of different kinds of music. Radio, live music, bands, singing groups. They played in the parks and there was a guy used to have a band right down the street from me and he rehearsed right in front of his building. Now that I think about it years later, I have reflected what I was hearing."

When he reached the sixth grade Craig wanted to play drums or clarinet in the school band, but those instruments were all grabbed up before it was his turn to choose and all that was left was a trombone. "When they gave me this big horn I said, 'Hey, all right! Let's go!'" Before two years had gone by, Craig was playing dances, weddings, and clubs with a local rhythm and blues combo four and five nights a week. One gig he especially remembers was in "a topless shake bar" that closed at 4 a.m.

"My parents didn't have too many problems with it," Craig assured me. "My mother said, 'Well, he's doing something positive, he's going to stay out of jail and he's going to finish high school.' The music pretty much kept me focused on getting away from those negative things that can happen in the so-called communities that I grew up in. People go to jail, some people die of overdoses, but the music took me into a lot of different circles and so I knew there were other things in life besides cutting school, standing on a corner, and robbing people."

Our final example of players on the scene as the idiom approaches the centennial of its birth convinces me that the music will flourish, as long as it is in the hands of such tradition-respecting, open-minded, and dedicated artists as twenty-four-year-old trumpeter Philip Harper, younger brother (by three years) of the drummer Winard Harper.

Philip Harper was born in Baltimore, Maryland, and grew up from the age of about seven, he told me, in Atlanta, Georgia. "My oldest brother Danny, who plays trumpet and piano, and Winard had a group together, kind of a pop group, all the way back in Baltimore," recalled Philip. "They used to play a lot and Winard got a lot of attention 'cause he was so young. And I wanted to do something and I tried it all. I tried sports and I went through all the instruments." Philip listed bass, guitar, and piano as among those he had tried by the time he was nine.

"Finally one day I just snuck into my brother's room and picked up my brother's horn and, you know, all of a sudden it just fit. I was nine years old. Danny was going to school at Clark College and he was beginning to buy a lot of jazz albums and he had some Lee Morgan albums. I went in his room one day while he was working on a Lee Morgan solo and heard it and that just blew my mind, it was like, 'Wow! Sure wish I could do something like that!' 'Cause the way Lee had the horn talkin', it just amazed me. I told my brother I wanted to play but he didn't believe me 'cause I went through so many other things. He just gave me a mouthpiece, 'Here, go and play on that for a little while,' and I did, I went everywhere, I buzzed on this mouthpiece, I went to school, buzzed on it, and finally I got suspended from school 'cause I was buzzin' on this mouthpiece. I guess my parents kind of figured that maybe I was serious about it, 'cause that was all I was doin'. I got my first trumpet when I was ten. They got it for me on my birthday."

When Philip was fifteen Winard graduated from high school and enrolled at Howard University in Washington, D.C. "During that time I was going back and forth a lot," Philip pointed out. "Every time I had a chance to get a vacation, then I would come up and spend that time with my brother. We had a lot of relatives there in D.C., cousins, aunts, my mother's mother lives right there in Baltimore. So we had people to stay with when we came up. It was like my last year of high school so I was between sixteen and seventeen."

I was curious as to what sort of support family had given to the Harper brothers. Did they offer encouragement and were there any musicians among the older members of the family?

"I guess it's just like any parents, if you see something that your kids are beginning to be successful with, then you're going to feel good about it and you will encourage them. Also, you know, they get those doses of reality where they have doubts about everything. My father was always trying to push my brothers when we moved to Atlanta because the rock band they had was doing okay in Baltimore.

So when we got to Atlanta he tried to set up a club for them to play at on a regular basis. It worked for a minute and then I began to play. I mean, we were all young then and in school and as things went on then it began to get more serious. We got the support we needed, but at the same time, like I said, every now and then you get those moments where it's kind of hard to see that that's really gonna be it.

"I have a cousin that played cornet and that's who, I guess, inspired my brother to play, and once he got into it, Winard wanted to play. When Danny would be practicing, Winard would be sitting down beating on little shoe-polish cans. And my father said, 'Maybe he wants to play drums.' So my father bought him his first drum set and that's how it got started for Winard. I guess he might have been about four or five years old then."

By the time that Philip began to spend his vacations visiting his brother in Washington, D.C., Winard had established himself to a considerable degree on the city's jazz circuit. Winard's regular partic- ipation in the weekend afternoon jam sessions led by pianist Lawrence Wheatley at the One Step Down had led to evening gigs in the rhythm section supporting horn players and pianists booked for the club's Friday and Saturday night action. He also played other clubs, includ- ing Mr. Y's, in combos made up of other young local musicians.

Philip occasionally sat in at the jams, and soon after moving to D.C. upon graduation from high school in Atlanta he and Winard took a group into the One Step Down for a weekend gig. I had been much impressed on a number of occasions with the exceptional talents of drummer Winard but was not prepared for the equally astonishing abilities of the eighteen-year-old Philip. The consensus among One Step regulars that evening—who, incidentally, were accustomed on the weekend nights to enjoying performances by the likes of Joanne Brackeen, Al Haig, Sheila Jordan, James Moody, Charlie Rouse, Woody Shaw, Phil Woods, et al.—was that here was a budding star, this wun- derkind horn player.

It wasn't long after that that the two brothers were offered schol- arships to study with Jackie McLean at Hartt School of Music in Hart- ford, Connecticut. They accepted but remained in the program for only a semester.

"I was glad to be there with Jackie," Philip assured me. "That was definitely a dream for me 'cause he played with all my favorite guys, and what time I could spend with Jackie was valuable. But it was kind of hard to be in school and maintain the studies plus try and work. It really left no time for practicing, so the stuff I *did* get to learn I couldn't incorporate. For me it just came to the point where I knew, 'I got to do somethin' about this.'

"Winard, he ended up with a gig with Dexter Gordon and he went off with him for a minute, and after that he decided, 'Okay, I guess it's time to go to New York.' So I said, 'Well, it's time for me to leave, too.' So we both went to New York and stayed with my mother's sister for about four months before we ended up working. Winard got the gig with Betty Carter and then about the same time I started working with Jimmy McGriff. For me that lasted about a year and a half. We never went overseas but we toured all through the States. Then that began to be just like school so I left, just went back to tryin' to practice a lot. Then Little Jimmy Scott heard me in New Jersey somewhere and hired me, which was like a dream for me 'cause I didn't know he was even alive at the time. I was just beginning to start listening to his records. I was pretty much off with him for the next two years up until the time I started working with Art Blakey."

Whenever Philip came off the road for a few weeks or so during the several years he worked with McGriff or Scott he regularly dropped around to the midnight jam sessions at the Blue Note. The trumpeter Ted Curson was leading the session, which ran until 4 a.m., six nights a week. He turned up so frequently that the management employed him to run the sessions during the summer months, which Curson spent in Finland, and on other occasions when the older trumpeter was not available.

"It was great for me, especially working every night," Philip explained, "probably one of my better learning experiences. I got a chance to hear and to work with all these guys who came to town. There was Ray Brown, Monty Alexander, Freddie Hubbard, George Benson, Slide Hampton, Paquito D'Rivera, Claudio Rodito, so many people."

In 1987 Philip joined Art Blakey's Jazz Messengers and remained in the group for about half a year. I was pleasantly surprised to catch him in performance with the combo at the Jazzfest Berlin that year and told him backstage after the set that his admirers back in Washington, D.C., were extremely proud of him.

Philip emphasized that he and Winard had kept a group of their own together during the five years or so that they were using New York as a base and touring, Winard with Betty Carter, he with Mc-Griff, Scott, and Blakey. They took part in a Blue Note club series of Young Lions and played gigs in New Jersey, Connecticut, Boston, and other locales. "We worked pretty much on a regular basis at a club in New York called Pat's, in Chelsea. We worked that like once a week, which kind of helped us get our thing together." Then in 1988 Poly-gram released the Harper Brothers' first album. A second release on the same label appeared in January 1990. In April of that year the Harper Brothers quintet was in first place on the jazz chart of *Bill-board* magazine.

"It's like, for the most part, all we do," Philip said. "We did a couple of festivals in Paris and we did the JVC festivals, Pori and North Sea." Philip had, independently from the group, already made the scene in

Japan several years ago with the drummer Motohiko Hino, brother
of the trumpeter Terumasa Hino, " 'Cause he came in the Blue Note
one night and sat in with me."

As Philip reached the end of his story, a rather full account for one
of so few years, it occurred to me that he might well have some interesting thoughts about coming up in New York in the 1980s. So I asked
him to reflect upon that, upon the present state of jazz there, and
upon his own circumstances.

"It's all I do," Philip began, "so I have to try my best to do it well
and do it better as time goes on. It's very hard 'cause this music is
hard, for one thing. You know, all my life I'll always be a student, I'll
have to study all my life. The more records I buy, the more humble
it makes me 'cause it just lets me know how much further I have to
go, and sometimes that seems so far.

"Then it's hard because the music is not where it should be. People
in the States take it for granted and we don't get the respect for this
music that we really should be getting. And financially it's not going
to make you rich. I don't really know any rich jazz musicians. That
can tend to be hard when you compare it to other things, say, maybe
some of your peers who they probably don't have to put out as much
and they're getting much, much more. But if your heart's there and
that's what you want to do, then you have to stick with it the best you
can.

"What I have seen," Philip said, assessing the state of the music and
the circumstances of musicians in New York in the 1980s, "before I
hit New York, that's when the improvement took place. For one thing,
Wynton Marsalis hit the scene and he opened the door for a lot of
young cats. They were able to say, 'Wow! Maybe there is some hope
for this music.' So what started happening was like a big flux of young
guys coming to New York. All of a sudden you have young cats coming from *everywhere*. And that's good and it's still happening. Every
time I leave and I come back, New York is like a whole new scene
because it's so many new cats here, and I mean cats who can play.

"I guess the not-so-good part is that there aren't as many venues to
incorporate some of these guys and help them get along with their
talent. Because when I got to New York there were like at least seven
or eight jam sessions you could go to all in one week. Sometimes in
one night you could make your rounds and hit about four or five
sessions and that gave me my chance to be seen and heard, and any
chance I could, I did. Both Winard and I, for those first four months
we got here, we hung out every night. I mean, even though we didn't
have any money, we did what we could.

"I was working on the street when I got here with this Japanese
drummer. We just put out one of the drum cases [for the kitty] and
played and we never really got stopped. Later they started to get hard
on musicians, requiring that you have a license, but we never really

experienced that problem out there. I would make quite a bit of money on the street, enough to go and hang out at night. And that's what I did, hung out every night until finally something came through. I made the sessions in New Jersey, I made the sessions in New York.

"Now it's almost none to make, so it's really kind of hard for the younger cats and the guys coming into New York because you don't have as many situations to be seen and heard in. There's no more Star Cafe, Barry Harris' club has closed down, and some of the other ones that was doing sessions, they're not there any more. The Blue Note is charging a cover for that session and it's really not a session. So it's kind of a sad situation.

"This music has so much and it's so bad and there's so many different places you can go just inside this music," Philip concluded, "I don't see why you need to even think about anything electric, there's so many things you can do. And the art of it is to try and find those things you *can* do, to always come up with something different every time you hit that bandstand."

It wasn't very long ago that, as far as the status of jazz in the larger society was concerned, the most appropriate observation was in terms of the adage, "Plus ça change, plus c'est la même chose"—the more things change, the more they remain the same. But the 1980s have indeed brought some changes vis-à-vis the status of the idiom. And while no great changes in the actual circumstances of the music and of the artists who create it have yet been wrought, at least one of those developments constitutes a beginning, a foundation upon which to build. At the very least, it is symbolic in terms of gradually changing attitudes in the musical "establishment" and in the greater society.

That the U.S. House of Representatives—and later the Senate as well—adopted in 1987 the Honorable John Conyers' resolution designating jazz an "American National Treasure" is a major step forward for the art form. In the congressman's words, "For the first time in the history of jazz we have the government acknowledging the art form." With the sense of the U.S. Congress on record as supporting such a concept it would seem that funding from both the public and private sectors should soon achieve a more equitable balance between the symphonic establishment and the jazz community. Not that such a turnabout will occur spontaneously. The jazz community must come together as it never has in the past and work together toward common goals.

Incidentally, the jazz community encompasses a much greater area today than it ever has in the past, and not simply in terms of that accumulation of styles referred to earlier. "The jazz umbrella, I think, is far broader than any umbrella in music," Bill Cosby pointed out to me in an interview that dealt with his jazz life, as a drummer and percussionist since his teens and as leader of his own recording and performing group in 1990. "I know that in classical music you have your baroque, avant-garde, American versus European, but I don't think that there's anything really like *our* umbrella. Our umbrella continues to get broader and broader, and if you go all the way back to when it first started, you can say that under that umbrella is just a group of people singing and clapping their hands and taking off on variations, and under it is Rufus Harley on bagpipes, Eddie Jefferson and singers who went 'oo-poo-pa-doop,' Ornette Coleman's 'Lonely Woman,' and something that I think the Europeans, more than Americans, do electronically, they just make sounds of crashes, playing sheets and sheets of sounds in freedom.

"The United States of America is a potpourri—I mean, you have *everything* in this country, there is no common force. So if you look at the music coming from these people—Caucasian, Asian, African, Native American—you've got all these blends. So that Ray Charles makes a hit album from country-western, so that Kenny Rogers does Marvin Gaye, so that Stephane Grappelli comes in to do Dave Brubeck. The blues—you got Chicago blues, Texas blues, Philadelphia blues, whatever. That umbrella."

Bill could have added a number of other idioms that now comfortably find shelter under his "umbrella," including Cajun, various Latin forms, r&b, and World Music and its various components. I asked Bill how he felt about including rap music under the umbrella.

"I think it's under the umbrella," he responded. "If you take the period of African drumming, which was a way of talking. I mean, that rhythm, you can't really deny the rhythm [of rap], because it is parade music, marching music, as far as I'm concerned. Then you take Eddie Jefferson scatting the notes. Well, these kids are not *singing*, but they do have some notes that they're *saying*. Like any soloist, it depends on what you're saying. These are today's messengers and *some* of the things that I've heard them say are absolutely brilliant. They're dealing with society. I think it's very, very important, and I think if it's well done, it's on a great intellectual level. And I certainly would welcome it."

For jazz to be recognized as a classical art form—which appears to be the case in some cultural circles here in the U.S.—is important for all jazz players, but perhaps especially so for black musicians and for black society.

"The thing that helped me most to develop my technique was listening to Clifford Brown," Wynton Marsalis revealed to me. "To me, he's *the* trumpet player—he and Louis Armstrong, and even Miles. They did things with the tongue and stuff like that. Just to try to learn a Clifford Brown solo. That's harder than anything you ever heard a classical trumpet player play," and in demonstration of this he rapid-fired an oral "tika-tika-tika—tika-tika" phrase. "But because the jazz musicians are black, they don't want to give the credit to them. See, that's the thing that outrages me the most is the fact that here are guys that have contributed a great body of music to American history and the only reason that they aren't being recognized for their labors is because the color of their skin is black or brown or whatever."

Saxophonist Donald Harrison, speaking from the jazz panel of a Congressional Black Caucus gathering in 1986, observed, "There is a great problem in America that children are not taught the same reverence for Charlie Parker as for Bach and for Beethoven." Donald's companion and co-leader of their combo, trumpeter Terence Blanchard, confirming his partner's statement, told of "asking children across the country if they listened to Dizzy Gillespie and they never heard of him."

Even more poignant was the story Steve Allen related to me. "A friend of mine was standing on Park Avenue recently and two well dressed little girls, one white and one black, came up wearing the same uniform of a private school in the area. I don't know whether they were all waiting for a bus or what the setting was but they struck up a conversation—he has two daughters of his own—and he discovered that the little black girl, about fourteen or fifteen and obviously upper-middle-class and not living in the slums, basically did not know anything about Louis Armstrong or Count Basie or Duke Ellington. These names meant nothing to her. She knew of Stevie Wonder and other folks of these days.

"Now it's always dangerous, from very limited evidence, to extrapolate philosophical points, but I wouldn't be surprised if that young lady has a lot of company, and that's a double-level pity. If she were *white*, it would be a pity because we're talking about three giant contributors to American popular culture and jazz, but if a well-educated young black child does not even know the names of those gentlemen, that's a problem of an additional order. So there's a lot of cultural things falling between the cracks in our society, but we keep slugging, doing what we can."

Of course, the concentration of the multi-billion-dollar recording industry upon the fast—and guaranteed—buck of million-plus-selling pop releases has played no small part in fostering ignorance of the jazz art among both youngsters and their elders. Steve had some thoughts on this subject: "I draw an analogy with book publishing," he confided. "No book of poetry ever makes big money either, unless it's second-rate poetry and sometimes that sells very well, but the best poetry probably sells about nine hundred and twelve copies. So the book publishers don't publish poetry for economic reasons, they publish it out of guilt or conscience or cultural obligation or however they may perceive it, and I think these record companies that are making more money than they know how to count from Stevie Wonder and Bruce Springsteen, for Christ's sake, ought to, just to be good people, push some jazz stuff, even knowing they're only going to break even on it. Can't they handle it after they made thirty million on something else? Of course they can, and they ought to be encouraged to do so, or *ordered* to do so, by their stockholders."

The present virtual absence of the jazz artist from the consciousness of young Americans was not always the norm.

"In the '40s, as teenagers, we *danced* to Tiny Bradshaw, Cab Calloway, Charlie Parker," Bill Cosby recalled for me. "We *danced* to those songs, so that 'Confirmation' was not only a song with a great solo, it was something to dance to. Miles Davis' 'Blue Haze' was a slo—o—w dance for us. 'Sister Sadie'—Horace Silver—we danced to. 'Night in Tunisia,' done by Art Blakey, we *danced* to that. We already had the rhythm. Those of us who are now in our fifties, we danced to things that, if you put it on for people now, they don't know what to do with it. But we had developed a dance that was all across the United States.

"So this music has meant something for me, deeply. When I was in college, living at home, every morning that I got up to go to class, I had a record player and I would hit the button and the record was already set to play, and it was Miles Davis and Gil Evans, Carnegie Hall, 'So What.' And I would get dressed to that, that started my day. I mean, that was my cup of coffee."

John Conyers echoed the foregoing account: "These were our cultural heroes, the jazz people," he told me. "Outside of sports it was the one thing a black person could look to with some pride." During the 1940s, when the future member of Congress was discovering the music and was beginning to learn to play the bass, "They were talking about a new music, a new expression, bebop. There were lively discussions and debates about its validity and how long it would last. The whole idea was you could concentrate on these jazz heroes."

Saxophonist Archie Shepp expanded upon this theme. "It's integral to my point of view," he pointed out to me, "that African-American music is the basis of the Negro's cultural experience here in the United States and that there are many social, political, and economic implications to this music."

In conclusion to this journey through nearly a century—there were, after all, recollections of Buddy Bolden included in our opening chapter—of jazz history, let us hear, briefly, the musicians' answers to these questions: Is jazz serious music? Does it deserve to be called a classical form? and Will it survive?

"Number one, it's a funny word," Maynard Ferguson responded to the first question. "Why should we be serious about something so joyful? Somebody who's in a chamber group has a tendency to act just like you'd expect them to act. Jazz music is such an adventure that it has to have that looseness, that joyful feeling, which is very appealing to young and old alike. Somebody once asked me, if I had stayed in classical music when I was young and made that my forte, would I not have been one of the world's great 'serious' trumpet players?"

Maynard chuckled before going on to the punchline. "And I said, 'Even if I had become the world's greatest classical trumpet player, I'd have still not been serious!' "

The second query I put to Richard Stoltzman, who has crossed back and forth between the classical and jazz worlds. "Oh, certainly," was the clarinetist's quick response. "Classical simply means something that has through a test of time become part of our accepted literature."

We're going to take the penultimate words of our book from an interview I taped for radio with drummer Art Blakey, through whose Jazz Messengers arguably have passed more players active in the 1980s than any other leader can lay claim to. "I just happened to live through it all," he proclaimed, "Dixieland, swing, modern jazz, and up to now— I guess because I'm so mean and evil—and I'm going to prove my point. I'm not out here for the money, I'm out here for the art form, and I think those who have gone before us should be given the credit that is due them. Duke Ellington, Louis Armstrong, Bix Beiderbecke, Art Tatum, Billie Holiday—these people have put down something, just like Bach and Beethoven, and it shall not perish." Art Blakey continued to tour the world with his Messengers until a few weeks before his death in 1990.

The final word comes from Bill Cosby: "I have this thing that I wrote," he offered as his assessment of the state, and status, of jazz in 1990. "It's a very short thing and I see it as animated. It's the end of the world, and winds are blowing, and it's very gray, no human beings alive. You don't see any bodies, you just see a lot of tombstones, and the camera pans, and there's paper blowing and all kinds of trash and empty tin cans, et cetera, et cetera. The camera pans and it comes up on this tombstone, and you begin to read very clearly—

JAZZ

and underneath, in quotes, it says,

"It Broke Even."

Index